Water in Times of Climate Change

Water in Times of Climate Change

a values-driven dialogue

Edited by Jan Jorrit Hasselaar | Elisabeth IJmker

AUP

The publication of this book is supported by the Van Coeverden Adriani Stichting and CLUE+

Cover design and lay-out: BVDT, Amsterdam

ISBN 978 94 6372 227 8
e-ISBN 978 90 4855 538 3
DOI 10.5117/9789463722278
NUR 900 | 940

Table of Contents

Introduction

PROF. RUARD GANZEVOORT

DEAN OF THE FACULTY OF RELIGION & THEOLOGY, VU AMSTERDAM, AND SENATOR IN THE DUTCH PARLIAMENT

Although the urgent COVID-19 pandemic now garners almost all political attention, the more stealthily emerging crisis of climate change will, ultimately, have a more structural and encompassing impact and is, therefore, crucial to the outcome of the Sustainable Development Goals (SDG) agenda. Water is and will be a key factor in how societies will experience, and be forced to address, the impact of climate change. But this is more than a technological challenge. It will affect the social and cultural fabric of our societies. For this reason, we must engage in a values-driven dialogue regarding water issues in times of climate

change. We must ask ourselves what is at stake and how we can tap into the richness of cultural and spiritual values to foster societal resilience.

Obviously, this will require the cooperation of actors beyond the scope of just governments, the corporate world, and technological innovators. Water-related challenges in times of climate change involve responses from environmental and climate scientists, scholars of religion and culture, social scientists, local and national governments and international organisations, financial corporations, technological business, NGOs, and religious and worldview communities.

With this in mind, the symposium 'Water in Times of Climate Change: A Values-Driven Dialogue' (Amsterdam, 6-7 November 2019) was designed as a new, collaborative way to address the urgency of water in times of climate change. The symposium aimed at building meaningful dialogues and a long-term alliance for a values-driven dialogue. More concretely, the symposium focused specifically on three major urban areas and involved actors from those contexts: Amsterdam, Cape Town, and Jakarta. In addition, the symposium 'Religion, Security and Peace: Religion and Belief in Contemporary Societies' (The Hague, 8 November 2019) critically explored the role of religion as a force for good in the public sphere. The book you are reading contains a number of the insightful and inspiring contributions to the symposia as well as subsequent reflections. Only a few months after the symposium, the world was confronted with the onset of the COVID-19 pandemic, making us all more aware that our world is deeply interconnected and vulnerable, and that the future is radically uncertain.

Water sensitive cities

The symposium addressed three significant challenges related to water and anthropogenic climate change from the interlocking dimensions of science, economy, government, and spirituality. Dialogues were intended to develop knowledge and find approaches that will marry technological advances with human and cultural values in order to envision feasible adaptations. In that sense, the symposium represented a holistic perspective in search of mutual understanding and shared languages for our common world. The three significant challenges were:

Threatening water. Paramount in the area of climate change and related issues are rising sea- and river levels, which pose a serious threat to many societies. The general sea level that threatens countries like the Netherlands and small islands in the Pacific Ocean, as well as disasters like tsunamis caused by earthquakes and volcanic activity, are a threat to the safety and viability of living environments. The risk of rising water levels will affect huge numbers of people in many of the metropoles in coastal areas. Increasing attention is now being given to disaster management and climate change adaptation (and possibly prevention). Less attention, however, is given to the existential fears related to this threat. 'Threatening water' plays a significant role in some religious traditions, for example in the stories of creation from the forces of chaos symbolised by the sea, or the story of Noah and the flood covering the earth. These stories may express the existential fear and offer modes of transforming fear into hope. The commonality of such stories in very different traditions is at odds with the lack of geopolitical consensus on dealing with these threats. The symposium explored how spiritual wisdom, geopolitical approaches, and technological issues and resolutions intersect.

Life-giving water. The dimension of water as a source of life is high on the agenda, gaining prominence due to increasing populations and the reduced availability of freshwater. Desertification and the increasing difficulty of providing populations with adequate water supplies pose serious challenges for metropoles and rural areas alike. One future consequence may be the significant increase of migration, which will also affect the areas with sufficient fresh water. Again, water as a source of life is prominent in most religious traditions, frequently occurring as a powerful symbol of life even under arid conditions. The tension between the fundamental need for water and the desired fulfilment provides a strong impetus for change. The symposium explored how economic solutions, political governance, and spiritual inspirations can reinforce one another.

Cleansing water. Clean water and sanitation on the SDG agenda relates primarily to issues of personal hygiene and health, but it also connects to sinking cities, the pollution of rivers and natural water supplies. Though governments are increasingly aware of the problematic consequences and the urgent need for improvements, there are many cultural and religious traditions that counteract positive behavioural change. Similarly, short-term economic interests can lead businesses to continue polluting processes. In religious traditions, the dimension of cleansing water links directly to the notion of purity / impurity that is intended to steer people's behaviour towards healthy and positive lifestyles. The symposium explored how spiritual and worldview values can foster sustainable change among populations and corporations.

In this book, these three themes are addressed from the vantage point of the three contexts. The theme 'cleansing water' will be approached primarily from the context of Jakarta, a rapidly growing metropole struggling with polluted rivers and overexploitation of deep groundwaters, leading to a sinking city and serious health risks. Amsterdam, the Dutch capital, around two metres below sea level, provides a setting for us to examine issues of 'threatening water'. The case study for 'life-giving water' is Cape Town, the South African city that looked set to be the first major metropolitan area in the world where the taps would run dry – so-called Day Zero.

Politics of hope

Water in times of climate change makes us aware that our world is deeply interconnected and vulnerable, and that the future is radically uncertain. This awareness can easily incite an atmosphere of fear and apocalypse, but it can also invite a politics of hope. In the midst of fear and self-interest, a politics of hope puts its trust in acts of empathy and solidarity, because it sees different or conflicting perspectives not as a source of polarisation, as they often are, but rather as a source of renewal.

Hope, then, is not a naïve invitation to a better world in the midst of radical uncertainty. Hope arises where the urgent reality of the world-as-is and the positive potential of the world-as-if converge. If we only look at reality, we can quickly become overwhelmed and disillusioned. If we only focus on the world-as-if, we can become deluded. But to combine those two antipodes is to create the energetic tension that moves the world.

The leaders of this world, be it in the spheres of religion, politics, academia, or business, are called upon to be agents of hope in envisioning how this world could be and engaging in this movement together.

Ecumenical Patriarch Bartholomew

JOHN CHRYSSAVGIS. THEOLOGICAL ADVISOR TO THE ECUMENICAL PATRIARCH ON ENVIRONMENTAL ISSUES | JAN JORRIT HASSELAAR. AMSTERDAM CENTRE FOR RELIGION & SUSTAINABLE DEVELOPMENT, VU AMSTERDAM

The VU Amsterdam and the Ecumenical Patriarchate have cooperated in the events on which this publication is based. From his enthronement in 1991, Ecumenical Patriarch Bartholomew, spiritual leader of the world's 300 million Orthodox Christians, has played a pioneering role in sustainable development. Admired and cited by Pope Francis in his encyclical ***Laudato Si'***, Patriarch Bartholomew proclaims that today's ecological issues require not only a technological solution, but also an articulation of the underlying ethical values and spiritual ethos; otherwise we would be dealing merely with symptoms.

Patriarch Bartholomew's many initiatives represent his conviction that environmental challenges must be resolved in dialogue and partnership with businesses, religions, governments, arts, financial corporations, media, NGOs, and academia. In June 1994, an ecological seminar convened at the school of Halki, the first of five successive annual summer seminars on aspects related to the environment and education (1994), ethics (1995), communications (1996), justice (1997), and poverty (1998). The unique seminars engaged leading scholars, environmentalists, civil servants, and students.

Convinced that environmental challenges must be resolved in dialogue and partnership with other religious faiths and scientific disciplines, in 1994 Patriarch Bartholomew established the Religious and Scientific Committee, which, to date, has hosted nine international, inter-religious, and interdisciplinary symposia reflecting on the fate of our seas, and forcing the pace of religious debate on climate change: in the Mediterranean (1995) and the Black Sea (1997); on the

Danube (1999) and the Adriatic (2002); in the Baltic Sea (2003) and on the Amazon (2006); on the Arctic Sea (2007) and the Mississippi River (2009); and, most recently, in the Saronic Islands of Greece (2018) – often under the joint auspices of the President of the European Commission or the Secretary-General of the United Nations. Drawing on the symbol of his sea-borne symposia, the Patriarch likes to say: 'We are all in the same boat.'

In 2012, the Ecumenical Patriarch decided to focus on smaller meetings, inviting scholars, specialists, and leaders to address targeted issues. The first Halki Summit (2012) discussed sustainability; the second (2015) explored literature and the arts; the third Halki Summit (2019) examined theological training; and the latest summit (2021) looked at life during a pandemic.

All of these endeavours have earned Patriarch Bartholomew several international awards, as well as the affectionate title 'Green Patriarch.'

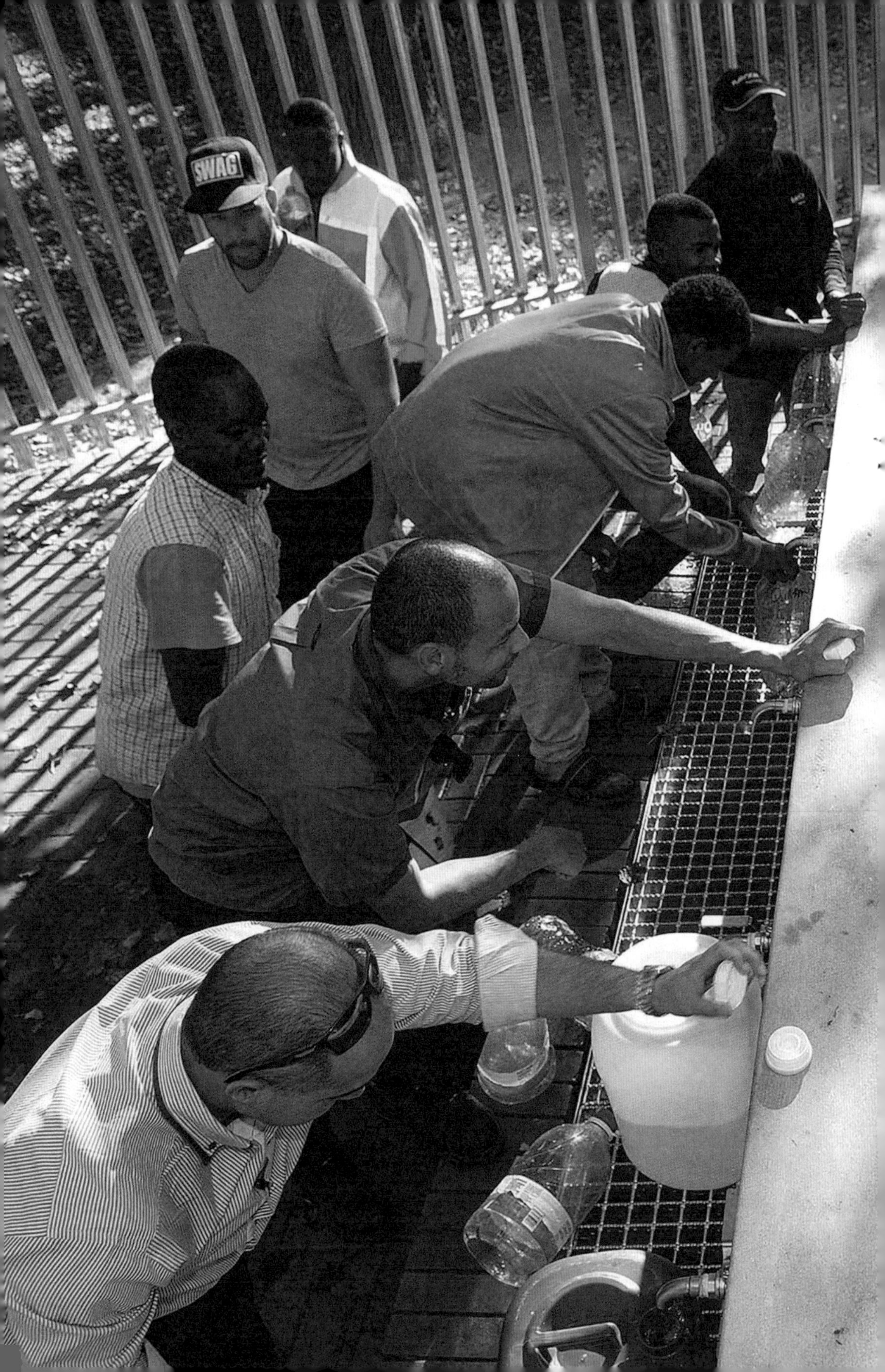
SWAG

1.

The Need for Dialogue

There is growing insight that issues related to water and climate change cannot generally be solved by one perspective or actor. For a politics of hope, partnerships are required (SDG 17) between different (academic) perspectives and between (inter) national government(s) and cities, academia, financial corporations, youth, arts, business, and religion. These partnerships, however, are not self-evident; even when partners share a sense of urgency, they often do not share the language to connect with one another. What would be a method to bring different perspectives and actors together in a meaningful way?

The contributions in this chapter come from academia, government, business, and religion. In his contribution from academia, Vinod Subramaniam states that we will only be able to begin to address the enormous challenges in these times of climate change by embracing the diversity of our differences. Caroline Nevejan, on behalf of the city of Amsterdam, argues for reintroducing ourselves to one another with respect, because we are all part of the same global community. From a business perspective, Jan Peter Balkenende accentuates that good dialogue contributes to making societies, communities, economies, companies, and organisations stronger. Ecumenical Patriarch Bartholomew points to 3 Hs as the ingredients of good dialogue: (1) humanism, with our own heart as starting point, learning to see with different eyes; (2) humility, reaching out to one another across our constructed barriers to create fuller understanding; and (3) hope, if we put our trust in a power that transcends our self-interest there is more than enough for all of us. This chapter also includes images by the renowned photojournalist Kadir Lohuizen, who recently completed a project on the global consequences of rising sea levels caused by climate change.

All contributions highlight the importance of the method of dialogue to bridge gaps between different perspectives and actors. Differences can be uncomfortable or even become a source of polarisation. However, all contributors consider differences primarily as a source of creativity and renewal.

An academic perspective

PROF. VINOD SUBRAMANIAM

RECTOR MAGNIFICUS OF VU AMSTERDAM

Since its inception in 1880, the Vrije Universiteit Amsterdam has cherished the principle of well-informed, incisive, but always respectful dialogue in our academic practice, with room and respect for different perspectives, interpretations, worldviews, and values.

Water is many things to the inhabitants of this planet. It is life-giving. It is cleansing. It is threatening. We take water for granted at our peril. The Dutch have some experience with this – as a low-lying country, below sea level, keeping water out and our feet dry is a national priority. Water plays an important role in virtually every religious tradition, and touches upon each of these characteristics – life-giving, cleansing, threatening. Allow me to reflect on the land of my birth, India, where the mighty rivers of the Indian subcontinent embody these characteristics.

Life-giving: The Indus, the Ganges, the Yamuna, the Brahmaputra, the Narmada, Godavari, and Cauvery rivers are lifelines for the teeming hordes, and essential for life in all aspects. Indeed, the life-giving nature of water also leads to bitter disputes, witness the decades-long dispute between the states of Karnataka and Tamil Nadu over sharing the Cauvery waters.

Cleansing: The Hindu tradition believes deeply in the cleansing aspect of water, and particularly of the great River Ganges.

Threatening: The mighty Brahmaputra regularly floods catastrophically. Much of this damage is of our own doing, due to rampant deforestation, resulting in increased siltation and soil erosion.

We will hear about all of these aspects, from different perspectives: religion, science, the youth – who will have to bear the burden of the choices we make today – the economy, governance. Bringing these different perspectives together is challenging, and perhaps uncomfortable, but it is only by embracing the diversity of our differences that we will be able to begin to address the enormous challenges that we face in these times of climate change.

To end, a short poem by the British poet Philip Larkin, entitled "Water."

Water

If I were called in
To construct a religion
I should make use of water.

Going to church
Would entail a fording
To dry, different clothes;

My liturgy would employ
Images of sousing,
A furious devout drench,

And I should raise in the east
A glass of water
Where any-angled light
Would congregate endlessly.

PHILIP LARKIN, 1954

A governmental perspective

PROF. CAROLINE NEVEJAN

CHIEF SCIENCE OFFICER, CITY OF AMSTERDAM

Amsterdam was born in a storm, in 1172, when the water of the River Amstel broke through to the IJ, flooding a small fishing village. When a dam was then built as a defence against the water, tradesmen immediately settled there, and Amsterdam started to develop, so the city archaeologists tell us. A few centuries later, low-lying areas around the city were drained for the construction of polders. Dykes and windmills kept the water out of the polder and when the tides were high, or the rains were severe, the people in the polder had to work together to survive. This resulted in a specific social structure, the so-called

polder mentality, in which everyone contributes every day to keeping the water out, and which disapproves of extreme emotions, because the water can come again at any moment and we will all need to work together again to survive. So the story is told.

Around the same time as the construction of the polders, the VOC, the Dutch East India company, was established. Trading goods by sailing to faraway countries, and creating amazing wealth for the Republic of the Netherlands, led to a different Dutch mentality. Driven by venture and profit, the VOC exploited the countries and the people they were doing business with. To our shame, the Dutch invented apartheid. We engage, in this volume, with three cities connected to this colonial history. In Cape Town and Jakarta, people suffered because of Dutch colonial power; in Amsterdam, this history is finally being acknowledged. In those times, advances in seafaring made these interactions possible.

We have arrived in a new era where we are starting to realise that we are all part of the global community that has to deal with climate change, and which demands that we reintroduce ourselves to each other with respect. Water is challenging us in new ways. From a Dutch perspective, this calls for a ⊠global polder mentality⊠, one in which we share knowledge, skills, and resources in order to take care of nature, water, and each other, and to abolish global and local exploitation.

As Chief Science Officer of Amsterdam, I value bringing together voices from academia and religious leaders, engaging with the students in our city. Insights from Cape Town and Jakarta will certainly help us here in Amsterdam to sharpen our thinking about the challenges of living with water.

VU
VRIJE
UNIVERSITEIT
AMSTERDAM

After us the deluge

KADIR VAN LOHUIZEN

AWARD-WINNING DUTCH PHOTOJOURNALIST

Kadir van Lohuizen's recent photo essay 'After Us The Deluge' shows the consequences of rising sea levels for mankind. He travelled to six different regions in the world (Greenland, US, Bangladesh, the Netherlands, UK, and the Pacific) and captured the effects of global warming. The following three photos are selected from this essay and reflect the challenges of water in times of climate change.

apg
accenture

An economic perspective

PROF. JAN PETER BALKENENDE

FORMER PRIME MINISTER OF THE NETHERLANDS, ERASMUS UNIVERSITY ROTTERDAM

In 1965, Dr Martin Luther King received an honorary degree from the VU Amsterdam. In a meeting with students at the Free University he said: 'I hope we can speed up that glad day when all of God's children – black men and white men, Protestants and Catholics, Muslims and Hindus, Jews and Gentiles – all of God's children will walk hand in hand, singing the old spiritual: Free at last, Free at last. Thank God Almighty, we are Free at last.' Dr King's remark was in line with the Four Freedoms of President Roosevelt: Freedom of Speech; Freedom of Worship; Freedom from Fear; and Freedom from Want. Freedom is always

connected with responsibility. Choosing for freedom implies also choosing for a responsible society, taking care of others, and the natural environment, now and tomorrow.

Raising the right questions about societal developments, about the importance of human dignity, solidarity, and stewardship is key. Today, in the world of the new economy we constantly talk about 3D. 3D, Internet of Things, big data, robotics are symbols of this new economic model. Here, I share my personal take on the 3D, those that are relevant for the quality of society: dignity, dialogue and development.

Dignity

Patriarch Bartholomew once said: 'The earth and humanity are created and intended to exist in a relationship of respect and harmony.' The Patriarch inspires us with this important remark. Indeed, reference was made to his views and insights in the famous Encyclical *Laudato Si'*. In *Laudato Si'*, the Pope talks very clearly about our Common Home and the necessity to think in terms of Integral Ecology. Across the globe, people have subscribed to the importance of this Encyclical. The Patriarch and the Pope have given very clear and meaningful messages. They inspire the whole world.

Inspiration can also be derived from the UN's Sustainable Development Goals. These goals are about ensuring quality of life for every human being, in the world of today and the world of tomorrow. It is encouraging to see that, amidst tendencies of nationalism, populism, protectionism, violence, hate, and terrorism, the SDGs offer a positive agenda for the world: global goals, global language, and global solutions. The views of the Patriarch, the Pope and the SDGs – they all contribute to dignity for everyone.

Dialogue

Dignity, solidarity, taking care of others and God's creation implies the willingness to engage in meaningful dialogues. Chapter V of *Laudato Si'* deals with the importance of dialogue: for international, national, and local politics, transparency in decision-making processes, we need dialogue between politics and the economy, religions in dialogue with sciences. Good dialogues make societies, communities, economies, companies, organisations stronger.

Some years ago, I was invited to participate in a conference in Thailand hosted by the Caux Round Table, an organisation that focuses on the implementation of 'Moral Capitalism'. It was heartening to see people with different spiritual backgrounds at that conference – Roman Catholics, Protestants, Muslims, Hindus, Buddhists, Humanists – reflecting together on the question of how these different traditions can contribute to the implementation of the SDGs. It was an extremely inspiring example of having the right dialogue.

In the business sector – where I am active, alongside my role at the university – we can see that those companies that are willing to integrate sustainability into their business models will choose for meaningful stakeholder dialogues. It is necessary to invest in good dialogues. SDG number 17 is about partnerships. It is clear that the UN underlines the importance of dialogue and partnerships to realise the other 16 SDGs.

In this country – the Low Countries by the sea – we are used to the necessity to work together in order to keep our feet dry. The essence of the 'Polder model' is dialogue and cooperation. This orientation has proved to be relevant, not only for issues regarding water management, but also in other domains: social agreements, energy agreements, climate agreements, pension agreements.

Development

Reflecting on dignity and having the right dialogue must have implications for development and must lead to concrete, practical results. Sometimes, we must act because of terrible circumstances. In 1953, the Netherlands was confronted with a flood that took the lives of 1836 people, including members of my family. What happened in 1953 had a huge impact on policymaking in the Netherlands – taking the right policy measures. In the year 2000, the world embarked on the Millennium Development Goals. Fifteen years later, it could be concluded that 70 or 80 per cent of these goals had been realised. The subsequent SDGs require concrete actions as well. Assessment tools are extremely important for analysing the progress we are making in the implementation of the SDGs.

It is good to see that universities, also those of applied sciences, are integrating the SDGs into their research and education programmes. Companies are rethinking their business models and business strategies. In the US, the Business Round Table published a 'Statement on the Purpose of a Corporation'. Following this statement, it is now a matter of execution. Cities are taking up their responsibilities with initiatives like U4SSC, United for Smart, Sustainable Cities.

So, a values-driven dialogue is necessary. But which values are we talking about? What are the practical consequences of these values? How can common strategies be developed?

The founder of the Vrije Universiteit – the university where I studied, defended my PhD thesis, and where I became Professor of Christian Social Thought on Economy and Society – Dr Abraham Kuyper, made a very important remark at the First Christian Social Congress in 1891, the year that also saw the publication of the Encyclical *Rerum Novarum*. Reflecting on the big societal issues of the nineteenth century, Kuyper said that an architectural critique of the structure of society was necessary. This remark is still relevant today. In fact, this is the message of His All-Holiness, of *Laudato Si'*, and of the SDGs.

A religious perspective

ECUMENICAL PATRIARCH BARTHOLOMEW

If there is one thing that we have learned from the global crisis and scarcity of clean water, it is that we cannot resolve the destruction of our planet or the depletion of its natural resources single-handedly. These problems clearly transcend national and political boundaries, just as they certainly exceed commercial and technological interests. They require working in cooperation, not venturing in isolation. They require building bridges. This is the message that we are called to accept and assimilate.

However, building bridges presents us with particular challenges as well as specific conditions:

Firstly, it requires an element of humanism. It means reaching inside the human heart, which is a bridge to the whole world. We must admit that we have failed in our vocation to care for God's creation. We have not done so deliberately; but whether through ignorance or indifference, whether as a result of selfishness or carelessness, we have not been ethical or faithful to our obligation to share the earth's resources with others. We have forgotten that water is God's gift to all people in every generation.

Secondly, it demands a sense of humility. It means reaching across barriers that we have created and zones where we feel complacent. We must accept that none of us can resolve these questions without working closely with other segments of society and disciplines of science. Politicians are called to listen to their constituents; and people are obliged to lean on their politicians. Corporations are supposed to protect the privileges of all; they cannot develop at the expense of others. Faith communities must pay attention to the warnings of the scientific community; and science must recognise the importance of mobilising the energy of religion.

Finally, it involves the dimension of hope. It means recognising that there is a power that transcends our limitations and a grace that fulfils our intentions. This factor is the unique contribution brought to the table by the Church. As the author of the Letter to the Hebrews (11:1-3) writes: 'Faith is the assurance of things hoped for, the conviction of things not seen [...] It is by faith that we understand that the world was created by the word of God.' Faith provides the hope that this world is larger than any one of us, that this world has more than enough for all of us, and that this world belongs to all of us as a precious gift from God.

These, then, are the challenges and conditions for building bridges: humanism, humility, and hope. In many ways, these are fundamental spiritual attributes and essential moral virtues. Because building bridges means building compassion and community. Building bridging signifies building stewardship and sustainability. And building bridges inspires faith and hope that there is a power that is greater than us – the word of God that patiently calls us to preserve creation and the hand of God that ultimately guides in the crisis that we face.

DN20
PN20

MONDAY
Cape Arg
SEPTEMBER 04
TOUGHER WATER CURBS
TO SUBSCRIBE TEL 0800 220 770

2.

Global Perspectives on Water and Climate Change

The contributions in this chapter deal with the need to address issues of water and climate change from multiple perspectives. Moving away from the dominant model, where we expect natural science and engineers to come up with the solutions to this supposed technical problem, the authors instead argue for bringing in new perspectives. Though someone from the religious scene might not be a usual guest at a conference on water, Patriarch Bartholomew challenges us to think about the moral and ethical dimensions of water issues in times of climate change. Since water serves as the source of all human, animal, and plant life in the world, having access to (clean) water is nothing less than a moral question. Answering moral and ethical questions requires partnership. The response from Bianca Nijhof illustrates just that. She describes how the Netherlands Water Partnership (NWP) creates a network of Dutch organisations in the water sector, bringing together different partners and stakeholders to better deal with questions around water.

Jeroen Aerts, an internationally recognised leader in the field of water and climate risk management, provides an academic perspective on the issues of water in times of climate change. While water is a source of life, it can also be a powerful force of destruction. Images allow us to see the impact of having too much water, floods, and the disaster that follows. Examples from the Netherlands and St Maarten demonstrate how water can also be a destructive force. Together, we must find a balance, between recognising that we all need water to sustain us, while remaining aware of our vulnerability to this force of nature.

Bringing these different perspectives together might be uncomfortable, but it is only by embracing the diversity of our differences that we will be able to begin to address the enormous challenges that we face in these times of climate change.

Closing session of Amsterdam International Water Week

ECUMENICAL PATRIARCH BARTHOLOMEW

How encouraging it is to have witnessed individuals and institutions aspiring to global cooperation and integrated solutions, as well as management and security related to a paramount issue of climate change. How inspiring it is to have witnessed individuals and institutions from all fields of life – from political and civil leaders to corporate and industrial innovators, but also from municipalities and cities to universities and NGOs – addressing water as the most precious resource of our planet.

In some ways, you may be wondering why a religious leader is participating at such a seemingly technical or technological event. Why would the spiritual leader of the Orthodox Churches throughout the world – indeed, why would our friend Cardinal Turkson, representing our dear brother Pope Francis as spiritual leader of the Roman Catholic Church throughout the world – be concerned about something so material and earthly?

The truth is that there is a close interdependence between the external environment and our internal environment. So, we feel at home among you because, by advocating the primacy of spiritual values in determining ecological action, we are proposing that there cannot be two ways of looking at the world. By underlining the sacredness of water as the blood that gives life to the world, we are highlighting that water can never be the property of any individual or industry. We have learned that those most harmed by global warming are the most vulnerable and marginalised, which means that the ecological crisis is directly related to advocating the right to water and to eliminating the scarcity of water. Therefore, clean water is nothing less than a moral crisis and a moral challenge.

The language of science and the enterprise of innovation, the determination of environmentalists and the initiative of policymakers – all of this energy complements the purpose of theology to open our eyes to the divine mystery and wonder of creation. All shades of human knowledge are completed by the ways of the heart, diverting us from a culture of exploitation and consumerism, while converting us to a culture of gratitude and generosity. In short, our efforts for water management and security, for water distribution and adaptation, are ultimately a cry of the human soul that recognises that water is essentially a gift to be treasured and shared with the entire creation. Our deep-seated conviction is that water is the inviolable and non-negotiable right of every human being and every living thing.

Water cradles us from our birth, sustains us in life, and heals us in sickness. It delights us in play, enlivens our spirit, purifies our body, and refreshes our mind. We share the miracle of water with the entire community of life. Indeed, each one of us is a microcosm of the oceans that sustain life. Every person in the world is in essence a miniature ocean. This is precisely why, over two decades ago, we declared that 'for human beings to contaminate the earth's waters [...] is tantamount to sin'!

It is the responsibility of us all – as state and religious leaders, as communities and individuals, as corporations and industries – to provide sustainable, clean, and safe water for the future of our cities and citizens, as well as all people and our planet. Such sustainability is not just sound technology, good business, or prudent politics. Such sustainability is the only way towards peaceful coexistence and global survival.

We owe it to our world and to future generations, just as we owe it to ourselves and to those who depend on us, not only to act but to act in partnership in order to sustain the critical resource and common heritage of water.

Bianca Nijhof

MANAGING DIRECTOR AT NETHERLANDS WATER PARTNERSHIP

The main theme of the Amsterdam International Water Week (AIWW) in 2019 was implementation of integrated solutions by cities, industries, utilities and financiers: from cases to bankable projects. This way we connect networks and create sustainable solutions across water, waste, urban development, energy, and finance. The AIWW sets the scene for new technologies, partnerships, research, and innovations to create the conditions for global cooperation and active progress. The presence, therefore, of His All-Holiness Ecumenical Patriarch Bartholomew at the closing session of the AIWW 2019 was a perfect example of how we can connect and collaborate across sectors and organisations. It illustrates how we make connections that far exceed the usual and predictable. After all, the value of water is omnipresent in every aspect of life, including the spiritual. By recognising this value of water, in meeting essential needs such as food and energy, we can make progress together across the Sustainable Development Goals.

As Patriarch Bartholomew pointed out in his speech, it is the responsibility of us all to provide sustainable, clean and safe water for the future of our cities and citizens, as well as all people and our planet. Besides my role as member of the Managing Board of the AIWW, in my daily work I am the Managing Director of the Netherlands Water Partnership (NWP). These are exactly the aims we strive for at NWP: a world in which nine billion people live well and in balance with the earth's resources with ample opportunities for future generations. We are on a mission to help solve global water issues. Patriarch Bartholomew called water 'the blood that gives life to the world' and we agree: ultimately water is in every one of us, as a source of life and as a connector of worlds.

Together with all participants at the AIWW and with all types of leaders in the world, both governmental and spiritual, we can make the future waterproof.

Global challenges on water and climate change

PROF. JEROEN AERTS

DIRECTOR OF THE INSTITUTE FOR ENVIRONMENTAL STUDIES, VU AMSTERDAM

According to the new UN World Water Development Report (2020), global water resources are under increasing pressure from rapidly growing demands and climate change. In their last assessment report, the Intergovernmental Panel on Climate Change (IPCC) states that climate change will increase the frequency and severity of natural hazards such as floods and droughts. Natural hazards already cause huge impacts to society with billions of dollars of damages and, on average, 60,000 casualties per year.[1]

Fresh water only constitutes three per cent of the earth's total water volume. Yet, these freshwater resources are critical for sustaining life. Population growth increases pollution of drinking water and increases demand for groundwater in arid areas. However, these groundwater resources are limited and, in many parts of the world, communities are faced with water scarcity and famine. The challenges to meet our food demand are huge and in only 30 years time (in 2050), we need to produce 70 per cent more food to feed 2.3 billion more people.[2] This means we must increase our yields with more advanced methods, but also become far more efficient with our limited freshwater resources required to grow crops.

Climate change will also increase flooding along rivers, in coastal areas because of sea-level rise, and in cities because of more extreme rainfall events. Moreover, even if we decrease our greenhouse gas emissions and limit global warming, an

1 Available from https://www.emdat.be/ [Accessed 10 March 2021]

2 "2050: A third more mouths to feed". Available from http://www.fao.org/news/story/en/item/35571/icode/ [Accessed 10 March 2021]

increased global population will simply mean more people living in dangerous, low-lying areas prone to flooding. In order to cope with these trends, we must adapt to the new future, on all levels of society. Governments will have to invest in flood protection, often with novel methods such as nature-based solutions. Think of beach nourishment to strengthen beaches against coastal erosion, the creation of green buffer zones in cities to store water, and the revitalisation of mangrove areas in coastal zones to protect communities against wave impacts.

Special attention is needed for the most vulnerable people, with the least capacity and resources to cope with climate change. Support is needed for these communities through joint efforts by the United Nations and other international organisations. It is in the interest of us all that we work together towards a more sustainable world.

JESUS

3.

Voices of the Next Generation: How Dare You?

"I should be back in school on the other side of the ocean. Yet you all come to us young people for hope. How dare you?!"
GRETA THUNBERG[3]

The voices of young people have become increasingly prominent in debates about climate change. Greta Thunberg gained international recognition for starting the school strike for climate, an international movement of school students skipping classes to participate in demonstrations demanding action to prevent climate change. She started alone, but quickly gained allies. Tapping into young people's anger and frustration, #FridaysForFuture climate strikes went global. In 2019, the movement had been a catalyst for 4500 strikes across over 150 countries.[4]

3 *The Guardian*. (2019) 'Greta Thunberg asks for lift back across Atlantic as climate meeting shifts to Madrid'. Available from: https://www.theguardian.com/environment/2019/nov/03/greta-thunberg-asks-for-lift-back-across-atlantic-as-climate-meeting-shifts-to-madrid [Accessed 18 March 2021]

4 *The Guardian*. (2019). 'Across the globe, millions join biggest climate protest ever'. Available from: https://www.theguardian.com/environment/2019/sep/21/across-the-globe-millions-join-biggest-climate-protest-ever. [Accessed 18 March 2021]

Whether it is organising a climate strike or carrying reusable bags, young people from around the world are taking actions – small and big – to protect their futures. For the first time in generations, young people are being told that their future will be worse than that of their parents: they know it is their future that is at stake. But why can't their future be full of promise and hope too? Young people often bring new energy, original ideas, and fresh perspectives. The untapped potential that young people can contribute to the debate around climate change increases the chance that we are missing out on unconventional and challenging thoughts and ideas that can help us shape the sustainable future we all need.

In this chapter, five voices from the next generation give their perspective on issues of water and climate change. Coming from China, Albania, Latin America, and the Netherlands, these writers share their worries but also their hopes, when looking to the future. In line with Greta Thunberg, they each describe steps that can or should be taken by the present generation to create a liveable future for the generations to come.

Water quality management

LIANG YU

I am a PhD student from China, a vast and ancient country that suffers simultaneously from both droughts and floods, as well as from serious issues related to the quality of both groundwater and surface water. Due to the increase of impervious surfaces and insufficient drainage systems, water logging has become a crucial matter for many big cities in China. More frequent and heavier rainfall events caused by climate change have no doubt aggravated this problem. Back in the summer of 2012, I was sitting in my rented house in Beijing, watching the rainwater turn into a big stream rushing around the house. That day, the deadly rainfall took over 70 lives. That was the trigger for me to become a hydrologist, to make my contribution to creating a better living environment.

As a young scientist, I have been studying urban water quality issues in cities like Amsterdam. Worldwide, water *quantity* management, the traditional aspect of water management, has always gained more attention and resources than water *quality* management. Indeed, water quality has more or less been considered a subjective concept. However, water quality is much more than an aesthetic and should not be overlooked by water managers. In many cities around the world, one of the most direct effects of climate change is the increase in the chances of eutrophication and algae blooms due rising temperatures and heatwaves such as the one experienced in Europe in 2018. This calls for unprecedented integrated water management scenarios to mitigate such stress on the urban environment. Game-changing actions that need to be taken may, to some extent, rely on the communication between scientists, governments, the public, and other parties. Take for example the concept of the "Sponge city" in China, demonstrating a collaboration between scientists and the government that we also see in relation to other concepts around the world such as water smart cities, sustainable drainage system, water sensitive cities, etc. I am happy to see the governments, scientists, banks, religious groups, and various parties of society gathering together in an event like the Water in the Times of Climate Change Symposium, connecting through brainstorming various perspectives, generously sharing experiences with each other to work on solutions to mitigate the effects of climate change. Development of such an integrated platform for the exchange of information and data related to climate change and water resources will no doubt stimulate and accelerate research and increase the communication efficiency among diverse parties across the world. And I wish to see water quality management gaining greater attention at a governmental level. As young scientists, as individuals, we should keep seeking inspiration and opportunities for a better environment in the future, even during this difficult time when the coronavirus is hovering over the whole world.

Caring for our water cycle

FRANCESKA TOMORI

PHD STUDENT OF ECONOMICS

For years now I have listened to the news, read in journals, and followed conferences about the phenomenon of climate change. We all know that water is a crucial component of our climate. Water moves between atmosphere, soil water, surface water, groundwater, and plants due to transfer processes such as evaporation, transpiration, precipitation, and run-off to complete its cycle. The water cycle is the continuous movement of water in the regions that are the focus of this volume. So, if water cannot circulate and complete its cycle correctly, there will be several side effects in these areas, and the climate will be subject to continuous change.

Unfortunately, we are facing this situation not only in Spain, where I have been studying and living for five years, but also in my home country, Albania. This leads me to believe that the climate is changing ever more rapidly, a direct result of human behaviour. Climate change is easily noticed in the seasons, as we experience shorter periods of autumn and spring weather. This is causing the ice in the polar regions to melt, thus increasing the oceans' water levels. Today, there are more natural disasters such as floods, scarcity of potable water, and even earthquakes. We are neglecting our care of the water cycle. We are not even conserving and saving water, as we cause continuous damage to our oceans, seas, lakes, rivers and every possible water source. I have witnessed the frequent unnecessary overuse of water. People leaving taps open, watering their gardens frequently, and letting water run for a long time. The result is that we are faced with water scarcity in many countries. It is now urgent to deal with this huge and worrisome problem. In my opinion, we are not doing enough to save the water and our planet from climate change. There is so much more to be done.

It is not easy to talk about water, although from my background, as a junior researcher in environmental economics and the economics of water, there is always a lot to say. All human and animal bodies are composed of a high percentage of water, making it the most important component of our life. Water also plays a central role in the world's economy. It is used in many industries, especially food production, agriculture, fishing, and others. It is used for washing, cooling, or heating different environments. Water allows for transportation by ships, while also serving as a means of entertainment. Hence, water is central to life, but also closely related to climate. All these benefits of water confirm the statement that: 'If there is no water, there is no life.'

Firstly, I would suggest to the youth of my home country, of Spain, indeed of all countries on the planet, that we must start solving this problem ourselves. Each of us is forms part of a future generation. Hence, we should all be held responsible for our actions. By reminding ourselves about the importance of water and climate, we can help prevent the overuse of water and the polluting of the environment. We should conserve and clean water, share our knowledge about water and environmental protection with the communities we live in, and also volunteer in climate and environment organisations to help our planet. Secondly, I call on media and environmental organisations to increase awareness of such problems. Finally, I am asking government authorities to contribute by engaging with different policies and investments. They should listen to the voice of active people, researchers, leaders of environmental organisations, etc. Through all these steps, I hope and believe that all of us can contribute to preserving water, improving the climate, and enjoying the valuable gifts that God has given to us through mother nature.

Water crisis and social inequalities

MONSERRAT VÁSQUEZ L

JUNIOR LECTURER AND RESEARCHER, ATHENA INSTITUTE, VU AMSTERDAM

I was born in Chile, a country that simultaneously has everything and nothing. A country with a variety of climates, fruits, vegetables, and landscapes, the largest mines of copper and lithium in the world, and the huge Pacific Ocean. But the only thing the Chilean people actually possess is hope. Almost the whole country was sold to private and foreign companies during periods of colonisation and dictatorship. But we, Chilean poor and middle-class people, believe that working together in communities and in solidarity will create a better future for new generations. We hope and work in solidarity because the Chilean state provides no social warranty; that is to say, in Chile, everything is in private hands. Even when we work 45 hours a week for a low salary (compared to other countries), we must keep working to ensure that we have access to education, proper healthcare, food, water, shelter, etc. Even the pension system is private, and a lifetime of work is no guarantee for a life of dignity after the age of 67.

The pandemic situation, together with climate change, is causing a huge crisis for the lower- and middle- class families in Chile. There is no social safety net, but the lockdown has meant that people are not allowed to work, and so they have lost their incomes and cannot afford to educate their children, to eat, or to access basic sanitation services. The World Health Organization (WHO) has stated that sanitation and proper hygiene are essential to contain the current pandemic. In a country where the agricultural sector accounts for 85-90 per cent of water use and is crucial to the economy, how can water be protected when it is a private resource? How can sanitation and access to drinkable water be ensured in a context of water scarcity and a global pandemic?

A root cause of the ongoing water crisis in Chile is the neoliberal policies and experiments that have been imposed since the dictatorship in the 80s. In this country, water management and control is in private hands and climate change is exacerbating the ongoing crisis. This translates into water being a privilege that private companies benefit from: more than 80 per cent of water management is directed towards mining, livestock, agriculture (mainly avocados and vineyards), and industrial production. In other words, water management is not focused on public consumption or viewed as a public right. Furthermore, there is no evidence that climate change has been taken into account in Chilean public policies, nor has there been an assessment of the private sector and its impact on water management. Both should be considered in policymaking.

Since October 2019, there have been massive protests against the systematic privatisation and social inequality, particularly evident in the National Constitution, that are the legacy of seventeen years of dictatorship. So, what hope is there for us? People helping each other and protesting against systemic exploitation as an effect of historical colonial processes, enhanced by the neoliberal economic model and its racialised and patriarchal politics. The systemic oppression Chilean people have experienced in a variety of ways, characterised by racial, gender, and class divisions rooted in colonisation, is furthered by the globalised neoliberal economic system. People are protesting against a system that is more worried about production and accumulation of private capital for a world's minority, rather than focusing on sustaining life. While long overdue, it is time to think ahead and apply new socio-economic models, such as degrowth and circular economy, with a renewed focus on local communities, eco-systemic equilibrium, and inter-species life rather than accumulation in the global circuit. In Chile, solidarity and empathy are helping people to survive this crisis with empowered women and indigenous peoples taking care of neighbourhoods and cooking common pots. This is our hope. Solidarity and empathy are our hope for a better and more just future for all beings.

Taking action

BAS ZAALBERG

MSC HYDROLOGY

I am annoyed by how we continuously shape and exhaust the earth for our needs. On the contrary, we should shape our needs to what the earth requires. If we care, it means there will be moments when we cannot take or get what we want. The act of shaping the world to our needs suggests that we understand the way the world turns. This is very arrogant as history has repeatedly shown that we have not mastered the earth's processes at all.

Many humans live their days without being overly conscious of diet, sufficient sleep, or other things that might help a malfunctioning immune system. Until they get sick. Then, suddenly, healthy diets, vitamins, and doctors' prescriptions appear! We see the same negligence when it comes to how we treat our planet. Recently, the first paper claiming that the Greenland Ice Sheet has melted past the point of no return appeared, implying an irreversible rise in sea levels.[5] In view of this, the planet seems to be sick. Yet, somehow we are ignoring the doctor's advice. Rather, our leaders seem to care more about preserving their power and positions, supported by forceful lobbies and motivated by personal benefits.

Take an everyday example of (in)efficiency: supermarkets. The tomatoes are from Spain, Dutch-caught shrimps are peeled in Morocco, then returned for sale in the Netherlands. Grains have a water footprint of 3000 litres per kilo, every egg equates to 135 litres, and your cotton shirt consumed 3500

5 King, Michalea D., et al. 'Dynamic ice loss from the Greenland Ice Sheet driven by sustained glacier retreat', *Communications Earth & Environment* 1.1 (2020): 1-7.

litres of water! Cow milk, sold at half the price of oat milk, has a larger water footprint and higher production costs. Meanwhile a minority of shareholders fill their pockets off the back of exhausting farm work. If we are to see progress, we must acknowledge that our problems are deeply integrated and complex. Local and fair trade products and efficient water (re)use may be a good starting point.

Thankfully, in my surroundings, I see an increasing number of (young) people who are active and convinced of the need to take care of our planet. More often it is seen as 'cool' to work or put effort into planet-preserving activities, like clean-ups, PhDs in relevant subjects, and initiatives or solutions through e.g. start-ups or NGOs. Nevertheless, the exhausting relationship we humans have created with the planet continues. We must find and encourage each other by connecting through networks and with the help of platforms for (the execution of) ideas. Blaming and pointing fingers is wasted energy, rather the stage needs to be cleared for some fresh and innovative ideas about living with our planet. Ultimately, the only ones to blame for not picking up our piece of the puzzle, step by step, day by day, will be ourselves.

Interdisciplinarity in solving global challenges

SAMIRA I. IBRAHIM

CANDIDATE 75TH LBB FOR INTERNATIONAL RELATIONS AT THE CLINGENDAEL INSTITUTE

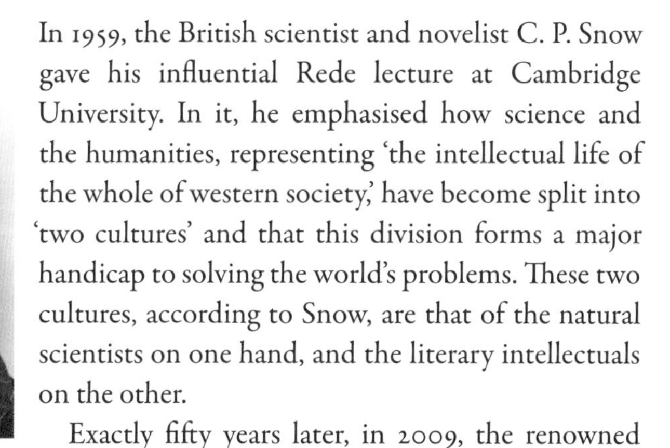

In 1959, the British scientist and novelist C. P. Snow gave his influential Rede lecture at Cambridge University. In it, he emphasised how science and the humanities, representing 'the intellectual life of the whole of western society,' have become split into 'two cultures' and that this division forms a major handicap to solving the world's problems. These two cultures, according to Snow, are that of the natural scientists on one hand, and the literary intellectuals on the other.

Exactly fifty years later, in 2009, the renowned American psychologist and emeritus professor at Harvard University, J. Kagan published his book *The Three Cultures*, in which he rejects Snow's proposed dichotomy of cultures, proposing instead a trichotomy of cultures that includes the culture of the social sciences, which he believed Snow had neglected. Thus, Kagan's main thesis is that academia is composed of three communities –natural sciences, social sciences, and humanities – each with its own culture of vocabulary, methods, and questions.

What, then, is the relevance of these two theses to the context of this book? On a personal level, I am intrigued by contemporary global issues and challenges, such as climate change, water scarcity, and the emergence of artificial intelligence. As a graduate student, I tend to approach these challenges from an academic perspective, which, for a long time, was shaped by my background in mathematics and engineering. However, recognising that I am also a religious and spiritual person with social and political interests, it seemed inadequate to approach these global challenges solely from the perspective of mathematics and engineering. To tackle this, in my perspective, incomplete approach to global challenges, I started following courses in politics, anthropology, and (Islamic) ethics at other universities, including the Azhar Institution and the American University in Cairo, alongside my regular studies at the Faculty of Engineering at Cairo University. Later, I moved back to the Netherlands to pursue a Master's degree in Water Management at the Technical University in Delft and further developed my interests in other academic disciplines at the Faculty of Religion and Theology at the Vrije Universiteit of Amsterdam. It was here that I came into contact with the economist and theologian Jan Jorrit Hasselaar, who invited me to deliver a speech at the 'Water in Times of Climate Change' symposium.

Whilst preparing for this event and reflecting on my own academic journey in understanding and studying water and climate change, I was reminded of Snow's work. Already more than 60 years ago, he pointed out the inability of natural sciences to solve global challenges such as climate change by themselves. And although interdisciplinarity and Challenge-Based Learning (CBL) are becoming more popular in several academic environments, by continuing to segregate natural sciences from social sciences and humanities at most of our (higher) educational institutions, we are leaving great potential for creative and effective solutions for contemporary challenges untapped.

SAMUDERA PASAI

4.

Voices from Those Who Dare

A politics of hope not only includes youth in responses to water and climate change. The current generation of their parents can also become an actor of hope by taking responsibility for our shared future. In this chapter, leading representatives from business, government, and the religious sphere respond to the challenge of the youth by addressing the question of water.

The authors argue that to fully understand water in all its complexity we must move beyond a narrative that addresses issues of water and climate change from a merely technocratic approach. Thankfully, there is a growing insight that we need perspectives from different sectors to develop a more comprehensive approach; one that embraces the value that each of these perspectives adds to the conversation. Water is many things to the inhabitants of this planet. It is life-giving. It is cleansing. It is threatening. By broadening our understanding of the issues at stake, together we can devise more creative and comprehensive solutions. We hope that this exercise contributes to greater awareness, structural exchange, and cooperation.

Moving from 'fighting against water' to 'building with nature' has had a tremendous impact on the work of Van Oord, one of the biggest dredging companies worldwide. Moreover, water is not just an ecological theme; it is an essential element of life and a profoundly spiritual symbol. Cardinal Turkson shows us how water has been used in religious traditions in many different ways, and what these traditions can teach us about looking at water from new perspectives. From a government point of view, both the UN and Hein Pieper argue that we must move beyond technocratic thinking, and instead need an approach that explicitly includes values and societal transformation. If we cannot make this step, we will lack the ability to properly address existential and moral dimensions, with the consequence that our responses to water in times of climate change will be less effective.

TRIKARYA

Building with nature

PIETER VAN OORD

CEO VAN OORD

About 30 years ago, I graduated from the Vrije Universiteit in economics. The science that focuses on material rather than moral value. Today, I represent a Dutch family-owned company with almost one hundred family members for whom values are extremely important.

First, I would like to address the central theme: water. It will not surprise you, but water has been the focus of marine contractor Van Oord for more than 150 years. Not from the perspective of cleansing water or life-giving water, but from the perspective of threatening water. The origin of Dutch marine engineering lies in the Netherlands' unique location on the North Sea and its eternal battle against the water.

Over the course of 150 years, Van Oord has grown into the company it is today: an international marine contractor specialising in dredging, offshore oil & gas infrastructure, and offshore wind. In addition, we are building solutions for challenges on the border of land and water.

The values of our company are We create, We care, We work together and We succeed. From a religious point of view, you could argue that the value 'We create' is quite ambitious. But if you look at Google Maps, you can see dozens of places on the earth that have been created by our company. Perhaps you have heard the saying: 'God created the world, but the Dutch created the Netherlands'.

Land reclamation creates space for urbanisation and economic activities. Projects aimed at increasing water-safety levels create value for the people living in coastal areas and polders and along rivers. The history of our company is entwined with the country's biggest marine engineering works, including the Delta Works and the Port of Rotterdam Maasvlakte II expansion. These were projects that fuelled prosperity and kept the population safe from flooding.

As a marine contractor, we manage and execute such projects. In the Netherlands, but also far beyond. Take Jakarta, a city where many water challenges converge, particularly threatening water. It is no coincidence that the Indonesian government is currently initiating serious studies on the relocation of the government centre, in part due to the threats posed by water.

This problem has a major impact on daily life in Jakarta. The city's people actually live with their feet in the water. It has become part of their everyday existence, because the capital of Indonesia is sinking – rapidly; about 10 to 25 centimetres a year in some places. If subsidence continues at this rate, parts of the city could be submerged entirely. A frightening scenario that is evocative for Dutch people, not least when we consider that Amsterdam is one of the lowest lying capitals in the world and close to the sea.

The reports of the Intergovernmental Panel on Climate Change (IPCC), point clearly to the global causes and consequences of climate change and rising sea levels. The latest forecast from the Dutch research institute Deltares indicates a potential one-metre rise in sea level in the Netherlands this century. So, the notion of threatening water is now acquiring a dramatic impact.

The urgency of climate change and rising sea levels means that companies such as Van Oord cannot wait for everyone to become sufficiently convinced. Our expertise is being called upon and we have a responsibility to act. I am convinced that a company like Van Oord is pre-eminently equipped to respond to this call. A guiding value for me, and for my family, is the idea that I did not inherit the company from the previous generation, rather, I have borrowed it from the next one. As a representative of the fourth generation of Van Oord, I am determined to be a good steward and pass on a responsible and sustainable company to the fifth generation. This stewardship also applies to the environment that we, as a company, are so closely connected with. After all, working with sand, water, and wind in the midst of society and marine ecosystems is our daily work.

During our 150th anniversary, in 2018, we presented a gift to society: we have made our expertise available to resolve the weak links in coastal defences on a global scale. And we have the ambition to play a leading role in climate adaptation in coastal areas. Together with the UN Global Center on Adaptation, we are finalising a joint Climate Adaptation Programme and we will soon start identifying the weak links. However, we cannot provide the solutions alone. The climate change challenge is simply too great. We must collaborate, especially with knowledge institutes, local partners, and financial institutions. And with you. You sometimes find allies in unexpected places. But we have to consider the character of the solutions as well: is 'fighting against water' still the right attitude and approach in all cases?

Climate adaptation confronts us with a dilemma. The pursuit of controlling water is a potential threat to the freedom of people and nature. Illustrative is the impact that coastal defence projects can have. At first glance, the enhanced water-safety levels they bring lead to more freedom. At the same time, these projects can have substantial adverse impacts on local communities and coastal ecosystems, threatening their freedom. Additional mitigation measures may then be necessary, but the tension between increased water safety and adverse impacts remains.

To manage this tension, I suggest replacing 'fighting the water' with 'building with nature'. There is no point fighting like a boxer against the forces of nature because we cannot win that fight. Rather, as subtle as a judoka who uses a hip throw, we should use the forces of nature to create the conditions for more sustainable solutions to ensure the well-being of both people and nature.

Building with nature appears to be effective for many projects, but we are not there yet. For our clients and the financial institutions that support them, it means that they must acquire confidence in building-with-nature solutions. As a maritime sector, we have the duty to provide the evidence and to facilitate the scale up. For Van Oord, this means that we must also invest in different

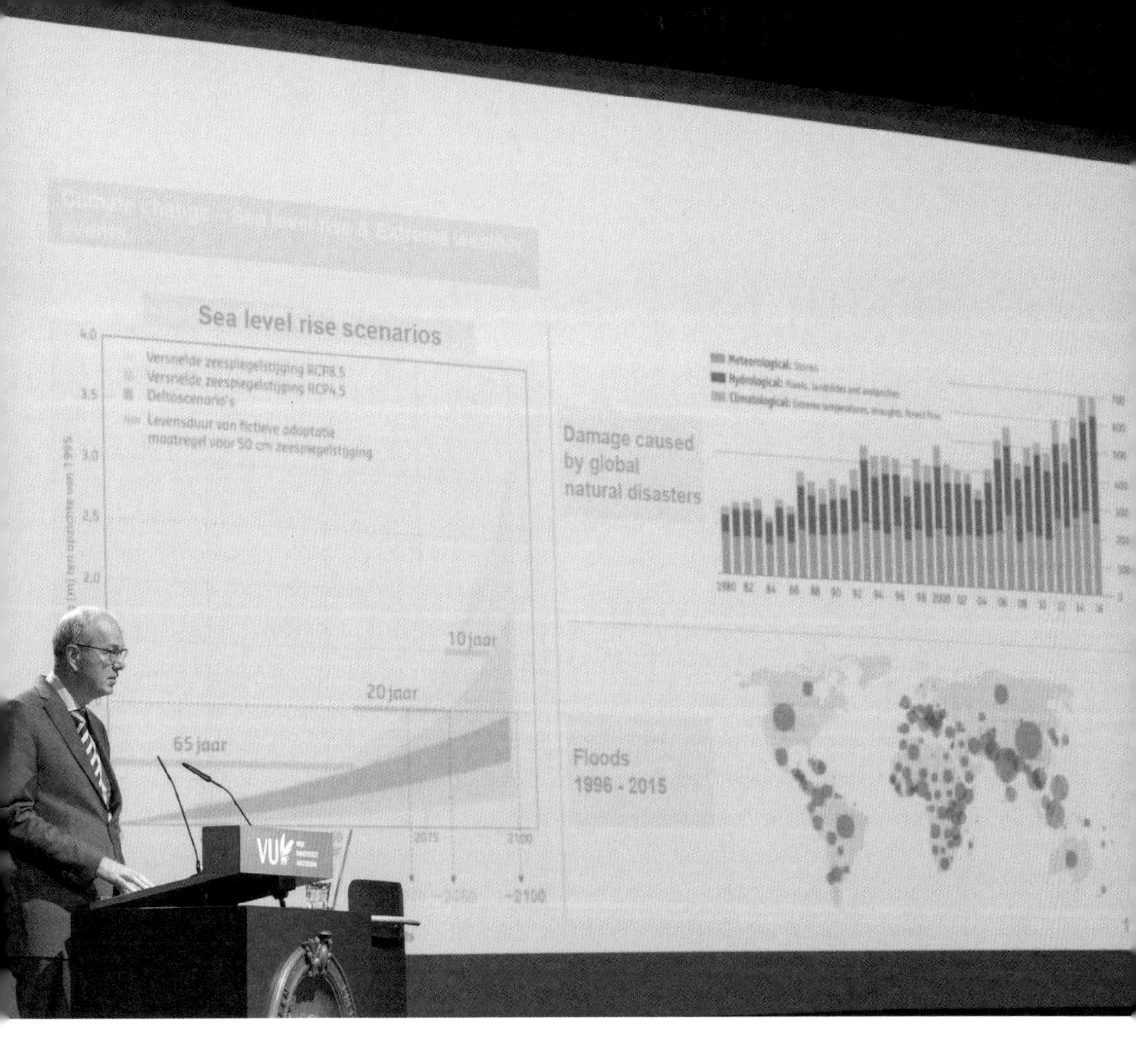

types of employees. Today, we not only work with hydraulic engineers, but also with environmental engineers, biologists, ecologists, and, perhaps in the future, specialists with a background in the humanities, too. We also notice that younger employees are far more familiar with building with nature than older employees. We must combine the enthusiasm of the younger employees with the experience of the seniors.

I want to conclude with a hopeful perspective. If we succeed in dealing with threatening water by replacing 'fighting against water' with 'building with nature', we will be taking a major step. One that perhaps comes close to what His All-Holiness Ecumenical Patriarch Bartholomew declared jointly with the late Pope John Paul II in their Common Declaration at the Fourth Ecological Symposium on the Adriatic Sea in 2002: 'It is not too late. God's world has incredible healing powers. Within a single generation, we could steer the earth toward our children's future. Let that generation start now.'

Faith for Earth Programme

IYAD ABUMOGHLI

PRINCIPAL COORDINATOR OF UNITED NATIONS ENVIRONMENT PROGRAMME'S FAITH FOR EARTH INITIATIVE

In John 4:10, Jesus said, 'If you knew the gift of God and who it is that asks you for a drink, you would have asked him and he would have given you living water.' In The Holy Quran, The Prophets Chapter 21 verse 30 states: 'We made from Water Every Thing Living.' Indeed, water is central to all living things, but also to virtually all development strategies, especially as part of the 2030 Development Agenda and the Sustainable Development Goals.

Water priorities and programmes feature in the strategic plans of many development institutions, public and private, secular and faith-inspired. However, there is a large disconnect between the calls to assure clean water and decent sanitation for all world citizens as a matter of rights and justice, and the realities of complex and often poorly harmonised policies and programmes on the ground.

- Due to human-induced climate change, polar ice caps are melting into oceans and mixing with salty water, causing a further loss of freshwater sources and sea-level rise.
- The discharge of the 80 per cent of all our wastewater, untreated, into rivers and lakes pollutes our waters.
- Some 800 children die each day due to preventable water and sanitation-related diseases.
- Around 2 billion people do not have access to a safely managed water systems.
- Floods and other water-related disasters account for 70 per cent of all deaths related to natural disasters.

These facts are just some of the facts that are the reality of the water situation today. Half of humanity now lives in cities and, within two decades, nearly 60 per cent of the world's population will be urban dwellers. Urban growth is most rapid in the developing world, where cities gain an average of 5 million residents every month. The exploding urban population creates unprecedented challenges, of which provision of water and sanitation is the most pressing and has painful consequences when lacking.

Two main challenges related to water are affecting the sustainability of human urban settlements: the lack of access to safe water and sanitation, and increasing water-related disasters such as floods and droughts. This leaves cities with the daunting challenge of managing the risks of increased flooding, a dearth of drinking water, and polluted water. Not only the poor suffer the consequences of these challenges, but they pay more to get their basic needs. For example, a slum dweller in Nairobi, Kenya, pays between five and seven times more for a litre of water than an average North American citizen.

In 2018, flooding affected more people than any other disaster type. In the past year, there were 315 natural disaster events recorded with 12,000 deaths, over 68 million people affected, and US$131.7 billion in economic losses around the world. Flooding not only affects human lives, but also the health and well-being of wildlife and livestock. This reduces the level of biodiversity, habitat potential, and food present in ecosystems, creating long-term impacts for surviving wildlife.

In August 2019, the UN Security Council met to discuss the growing new threats to world peace and security. Around 80 countries jointly agreed that the greatest impending threats to humanity were likely to be triggered not by terrorism, nuclear war or conflicts around the world, but by climate change. Climate change migration is one of the most challenging issues facing humanity, not in the future, but now. The threat of rising sea levels, caused by climate change, could result in a new category of 'environmental migrants' escaping from their sinking homes to neighbouring countries. The increased frequency, severity, and magnitude of extreme weather events all over the world will continue to generate humanitarian crises. Fluctuations between too much water and too little will have significant political, social, environmental, and economic consequences. If not managed effectively in a fair and inclusive manner, water can be a conflict driver.

Despite the complexity of the challenges, water is also a resource for collaboration. While the past 50 years have seen around 40 acute violent water conflicts, some 150 water treaties were signed around the world. An integrated approach to addressing the nexus of climate change, water management, and cities, is essential to fully account for the social, economic, political, and security impacts at both the local and the global levels.

To implement the 2030 Agenda and the Sustainable Development Goals, the United Nations Environment Programme (UNEP) and other UN agencies are providing advisory service packages, at global, regional, and national levels, on water management strategy. UNEP has developed strategies and taken on projects with global multi-stakeholder partnerships comprising governments, intergovernmental agencies, academia, the private sector, and civil society. UNEP is advancing the Integrated Water Resources Management, a process promoting the coordinated development and management of water, land, and related resources to maximise economic and social welfare in an equitable manner without compromising the sustainability of vital ecosystems. For example, Indonesia's Java island, seat of the capital Jakarta, accounts for 57.5 per cent of the country's population but has only 4.2 per cent of the country's water resources. Moreover, all six of its rivers are heavily polluted. UNEP has been working with the authorities and stakeholders to improve the poor Integrated Water Resources Management (IWRM) in Indonesia, where the severest obstacle to water management in the country is financing, in the interim, while capacity building and institutional frameworks are long-term concerns. The UNEP is implementing Flood and Drought Management Tools in the Wadi El Ku Catchment Management Project in Sudan to address water-related conflicts and disasters. It has become clear that only through dialogue can water be well managed, secure peace, and strengthen identities.

Practically, religions and culture can address climate change, biodiversity loss, pollution, desertification, and unsustainable land and water use by promoting a positive behaviour change. 84 per cent of the global population associates with a religion. UNEP's Faith for Earth Initiative is a strategy to mobilise faith-based organisations (FBOs), communities, individuals, and other entities through advocacy with religions on environmental issues. Our aim is to strengthen partnerships with FBO leadership for policy impact, to green FBO assets, and to transform the financing of the Sustainable Development Goals while presenting science-faith-based evidence. We are establishing the first ever global coalition of faith leaders on global environmental issues. This coalition will not only include eminent faith leaders, but also the young men and women who will become the leaders of tomorrow. A council for young faith leaders will be able to mobilise masses of youth to advocate for and engage in finding solutions to the global environmental crisis.

UNEP has been engaging with young people over the past decade. In September 2019, youth leaders from more than 140 countries and territories were invited to a Climate Action Summit to share their solutions on the global stage, and to deliver the clear message to world leaders that we must act now to address climate change. With direction and support from His Holiness Cardinal Turkson, in July 2019, UNEP hosted the 2nd International Conference of the 4th anniversary of *Laudato Si'*, with the participation of more than 300 young faith leaders who committed to engage at the global, regional, and local levels. The UN's ActNow Campaign is a call for individual effort. Tackling the climate emergency requires effort from all sectors of society. Also in September 2019, seven young environmental champion prize winners were endorsed by UNEP for their astonishing contributions to sustainable development progress. Twenty-one-year-old Brazilian Anna-Luisa Beserra, who designed a cistern solution for clean water, won the prestigious Young Champions of the Earth Prize in Latin America and the Caribbean. Molly Burhan was the first faith leader to win the prize for mapping land owned by the Catholic Church to measure its carbon footprint.

Mobilising partnerships is essential for tackling water issues and climate change. Engaging and partnering with faith-based organisations is a vital way of capitalising on the cultural diversity and the ethical values of religions. The Faith for Earth Initiative is actively committed to ensuring that the sound stewardship of natural resources is a fundamental human value and responsibility.

Water and religion

CARDINAL TURKSON

ROMAN CATHOLIC CHURCH

From the point of view of the Holy See, as a Church institution and a faith group, water is not just an ecological theme. It is an essential element of life and a profoundly polyvalent spiritual symbol. Though intimately related, the two values require particular attention and studies to precisely reveal the nature of their relationship.

Thus, beginning with a consideration of the wealth of symbolic values and applications of water in religions, but especially in Christianity, we shall proceed to consider, from the point of view of the Holy See, its ecological significance for the life of the human person and his/her world: its great necessity for life, and its presence and availability to the human family and its 'garden home', the earth. The necessity of water for life leads to the Church's description of a 'right to water'. The latter case of the presence and availability of water for all and for the earth have resulted in the Church's describing water as a 'common good' and a 'common patrimony' of the human family and of the earth.

All of this must now be considered in the context of the phenomenon of climate change. Climate change is believed to manifest itself most patently, and its effects most felt, in changes in the water cycle, namely, in the delicate balance between evaporation and precipitation. Indeed, 'water is the primary medium through which climate change influences the earth's ecosystems and, therefore, people's livelihoods and well-being.' It is, therefore, about water, its availability or its lack, and the weather. 'Weather' is the atmospheric condition, based on the sun, of a place at a particular time. When this weather is studied over a long period, it describes a 'pattern' that is called 'climate'. As the climate changes, so the atmospheric conditions also change, causing temperatures to rise or fall. The rise in temperatures causes droughts (water insecurity and unsafe water), storms and floods, melting glaciers, rising sea levels, etc.; falling temperatures cause freezing conditions.

Water as a source of life on earth and religious symbolism

In many religious traditions, including Christianity, water is a primordial element, and it is believed to be the *fons vitae*, the very source of life. The Scriptures of biblical religions (Judaism and Christianity) begin with the narrative of the creation of the world, where the hovering of the Spirit of Elohim (God) over the waters (Gen 1:1-2) sets the scene for the creation of all that exists (cf. Gen 1:6-7, 9-10, 20-21).

Today's science confirms this to be true, namely, that at the origin of all things is water. Accordingly, what makes the earth unique in the known cosmos is the presence of water on it: and water as the source of all life. But water only exists on earth because the planet is positioned within a specific narrow band around its mother star, known as the Goldilocks Zone. This position of the globe at an optimal distance from the Sun makes it possible to have water on its surface and, thus, for the garden home of humanity and its oceans to be the womb of life on the planet. In this regard, not only do scientists affirm the origins of life in water, they also claim that life is possible only if the availability and presence of water is continuous.

Water covers over two thirds of the surface of our planet; and an almost equivalent proportion of the body mass of every adult human is composed of water. So, naturally, water is the most potent symbol of life.

Water as source of human and divine life

For Christians, water is an element that communicates not just ordinary life, but God's presence and divine grace. On account of its natural cleansing powers, water is used in several religious rites of purification and atonement. It is also an essential element of several rituals that promise and ensure reconciliation and wholesomeness of the human body, of human life and of human relationships.

It is used in several Christian and traditional rituals of initiation (*rites de passage*) to mark a passage to maturity and adulthood, or, in the case of the Christian rite of baptism, that symbolises the passage from death to life. This Christian initiation rite is prefigured in the biblical story of Ancient Israel's crossing through the waters of the Red Sea to mark its deliverance from Egyptian bondage for a life of freedom, as God's people (Ex 13-14). In its current celebration as baptism or a Christian rite of initiation, the waters of the Red Sea are replaced wit baptismal waters. When someone is immersed in this water, or when it is poured onto someone, their old way of life (of sin or living for oneself) dies in order to embrace a new life of living in God's love and for mutual service (well-being) (cf. 1Pt 3:20). Because of the fundamental significance of Baptism as a rite of purification and a rite of passage to becoming a member of a religious group, the Catholic Church, for example, precedes gatherings to celebrate the life of communion of its members in the Eucharist with a reminder of their baptism: their washing by water and the Holy spirit, in a rite of sprinkling with water called the 'asperges'. Referencing biblical passages such as Ezek 36:25ff., Titus 3:5, and Hebrews 10:22, the rite of 'asperges' recalls baptism, when the faithful were purified for a new life with their Lord in the Eucharist.

Finally, on account of its natural cleansing power, water is blessed, in turn, used to bless and purify objects and persons that are dedicated to deities and intended for divine purposes. As a corollary, what is set aside for divine purpose also enjoys a special divine protection.

In this sense, water is used in many popular rites at shrines and places of pilgrimage, such as in Lourdes, in France, as a source of healing.

Water as a source of illness, violence, and death

If water is the source of life for all of our planet's creatures and human beings, then the pollution and contamination of water due to agro-chemicals, cultural habits related with poor sanitation, extractive industries, and other industrial uses must be considered sinful (cf. *Laudato Si'*, 8). Indeed, the lack of access to fresh and clean water is one of the main causes of death and disease. According to statistics, one in every two sick people in the world suffers from symptoms linked to a lack of good drinking water or from the use of contaminated water. Around 80 per cent of diseases in the developing world and one in three deaths there are linked to the use of polluted water. More children are killed by dirty water than by war, malaria, HIV/AIDS, and traffic accidents combined. Every eight seconds, a child dies from drinking dirty water.

Water scarcity is increasingly becoming a source of serious conflict. A tendency to consider water as a marketable commodity rather than a common good and a common endowment of our common home, the earth, means that our water management strategies are adversely affecting the poor. Access to water is a source of global inequality. (cf. *Laudato Si'*, 28-29, 48).

Syria is a case in point: the country suffered its worst drought in history and, consequently, a record crop failure in the period 2006-2010. As a result, 1.5 million people in Syria have emigrated from rural farming communities to urban areas. While this was not the cause of the civil war, it certainly fanned the flames.

Another case in point is the Sahel region of West Africa. Here, droughts provoke the incessant conflicts between pastoralists and farming communities. The pastoralists are predominantly Fulani Muslims and the agricultural communities predominantly animist and Christian. Consequently, the conflicts tend to assume a religious character, fomenting religious tensions in communities that are all too readily exploited by arms traffickers.

Similarly motivated are the systematic threats to and the killings of environmental activists in Latin America. At a recent synod on the Amazon, bishops and leaders of indigenous communities frequently spoke of the systematic killing of environmental activists: the avowed carers of our common home! The emerging and rising tensions between environmental caretakers and 'developers' of natural resources in the region point to conflicts that are destined to become more acrimonious and frequent as the fall-out of climate change becomes increasingly severe.

Spirituality and caring for water

In order to address the issue of water being corrupted, from a source of life into a source of conflict and death, Pope Francis invites us all to join forces and work together for the radical change of an ecological conversion. This change is urgently needed in policy formulations, laws, and regulations, but most importantly it is vital if we are to cultivate sound ecological virtues (cf. *Laudato Si'*, 211). Such virtues will change habits and develop an attitude of caring for the earth's resources, especially water! We must be prudent in our use and disposal of water, and ensure that everybody has access to this vital natural resource. In fact, the right to water is crucial to the pursuit of other rights, because without water there is no life. According to the Catholic social tradition, the obligations regarding the right to water extend across space (the entire world and human community) and time (for future generations).

In short, we need specific persons and communities to educate others in 'ecological citizenship': an education system that creates a 'bold cultural revolution' (*Laudato Si'*, 114); one that helps reduce the socio-ecological debt created by the disparity of access to clean water (cf. *Laudato Si'*, 30), which is also symptomatic of the disparities in access to food, health, and a good life (cf. *Laudato Si'*, 31).

Religious communities have a special responsibility in this regard. In the case of the Catholic Church, our use of water in the sacraments (God's privileged presence in our life) can help us re-value water as a source of life. Water unites us, it is the source of life – for the earth, for creatures, and for humans. But water is also the symbol of eternal life. To leave millions of sisters and brothers and some ecosystems without access to water is not God's intention; it is a sin. Conversely, to care for water and to ensure its availability, as a universal good that is shared amongst us all, is to echo God's invitation: 'Come to the water all you who are thirsty; though you have no money, come!' (Is 55:1).

Such a frame of mind and heart fosters a profound interest in caring for water; and it certainly constitutes a spiritual inspiration for the moral stance that the Church takes in its teachings on the matter and in its interventions at international meetings.

The Holy See and World Summits & Discussions on Water

A) The right to Water, as a Common good:
Pope Pius XI taught the Church and the world that, 'created things may serve the needs of mankind in fixed and stable order.' Accordingly, in 1961, when Pope John XXIII was dealing with the problem of the drift of rural and agricultural population into cities, he proposed a solution that included the provision to ensure 'drinking water' in rural areas. A year later, Pope John XXIII convoked the Vatican Council II (1962). He did not live to conclude the Council, but its document on the role of the Church in the Modern World, 'Gaudium et Spes', deals explicitly with the theme of life (indeed, it is mentioned 200 times) and the place and dignity of the human person in the world: "[...] all things on earth should be related to man as their center and crown."

In 1965, St. Pope John XXIII's successor, Pope Paul VI, concluded the Vatican Council II. However, in his previous travels to Latin America (1960) and to Africa (1962), as Archbishop of Milan, and later, to India, as Pope, St. Pope Paul VI identified the issue of 'underdevelopment' as the great 'social question' of his times; an issue that his Encyclical Letter, Populorum Progressio (1967) sought to address. In a brief section of this work, entitled 'Issues and principles', Pope Paul VI stated: 'if the earth truly was created to provide man with the necessities of life and the tools for his own progress, it follows that every man has the right to glean what he needs from the earth.' Here, water is not explicitly mentioned, but the Pope enunciated a principle that enshrines man's right to the goods of creation, including water. Accordingly, Pope Paul VI not only taught us that: '[...] under the leadership of justice and in the company of charity, created goods should flow fairly to all.' He also called for institutions to teach about the relationship between man and creation /nature for man's true development; for, as he observed later (1971), man is also capable of abusing the created goods that fall to him by right: 'Man is suddenly becoming aware that by an ill-considered exploitation of nature he risks destroying it and becoming in his turn the victim of this degradation.' As we have seen, this certainly includes man's use of water. Paul VI also called for the 'the establishment of a world authority capable of taking effective action on the juridical and political planes' to ensure that the goods of creation ensure man's true development.

As if the United Nations Organization was listening, in 1972 the Stockholm conference launched the UN Program on the Environment (UNEP). Just prior to the launch, Pope Paul convoked the Synod on Justice in the World, in November 1972, where this crucial observation was made: '[...] people are beginning to grasp a new and more radical dimension of unity; for they perceive that their resources, as well as the precious treasures of air and water – without which there cannot be life – and the small delicate biosphere of the whole complex of all life on earth, are not infinite, but on the contrary must be saved and preserved as a unique patrimony belonging to all human beings.'

After the very brief pontificate of Pope John Paul I, Pope John Paul II succeeded Pope Paul VI, taking and developing further his teaching on the place of humans in the created world. The thrust of Pope John Paul II's teaching was to extend a moral structure to the ecological question, namely, the relationship between the person and his/her environment (the created world); and these are his considerations: Since 'one cannot use with impunity the different categories of beings, whether living or inanimate – animals, plants, the natural elements [...], one must take into account the nature of each being and of its mutual connection in an ordered system, which is precisely the cosmos.' Secondly, natural resources are limited, and not all are renewable. If we treat them as inexhaustible and use them with absolute dominion, then we seriously endanger their availability in our own time and, above all, for future generations.

Accordingly, humans should not fall into the anthropocentric error of exercising absolute dominion over created goods, because of their capacity to transform, and, in a certain sense, re-create the world through talent and work. Rather, the things that God has created are for our use, to be employed in a responsible way, for man is not the master but the steward of creation. Similarly, when any created goods fall to a person, as a result of the right to private ownership, their use must always be subordinated freely to their original common destination as created goods.

Thus, between the Pope who opened the Vatican Council II and the Pope who closed it, much prominence was given to teaching about access to created goods (including water), as universal goods that rightfully belong to members of the human family. Pope John Paul II deepened this teaching in the context of the relationship between man and his created world: between a human ecology and a natural ecology. In so doing, Pope John Paul II gave a moral structure to man's right to access and use created goods. From this, the teaching of humanity's right to access created goods will become an official and a universal Church teaching in the Compendium of the Social Doctrine of the Church.

Under the pontificate of his successor, Pope Benedict XVI, the study of the relationship between the two ecologies of man and of nature (created world) deepened.

For Pope Benedict XVI, not only does the ecological question (relationship between man and his environment) have a moral structure, it is essentially dependent on and determined by the moral tenor of society. Furthermore, man not only has a right to access the endowments of nature (created goods), he/she also has duties towards them and the environment. Thus, man and is inextricably linked to the world. They are united by a reciprocal relationship, leading Pope Benedict XVI to describe man and his environment as a Book of Nature that is one and indivisible. The human and the natural are bound by a 'covenant relationship', which should mirror the creative love of God; for, the duties of human beings towards the environment and created goods flow from their duties towards each other. The relationship between humans and their environment is deeply ethical and moral is, in the sense of both rights and duties. Thus, humans not only have a right to access and use created goods, like water, but 'matter is not just raw material to be shaped at will; rather, the earth has a dignity of its own and we must follow its directives [...] We must listen to the language of nature and we must answer accordingly.' We know that the human family has not been particularly good at respecting the dignity of nature; and the description of climate change as 'anthropogenic' (originating in human activity) may also entail moral misconduct on the part of humanity.

In Pope Francis, the blend of the rich heritage of the teachings of his predecessors on the environment (created goods) and his pastoral experience in the Regional South American Church of the Aparecida Document fashioned a witness and a strong advocate for the covenant that must exist between the environment (nature) and its dwellers. He has expressed this as the need to care for creation, as integral human development, as concern for the poor and the aged, and as a call to listen to the cry of the earth and of the poor, in his homilies, addresses, and messages to various audiences at and events, as well as in his Apostolic Exhortation. This culminated in the writing of an encyclical on natural and human ecology, *Laudato Si'*.

In *Laudato Si'*, Pope Francis says: 'When we speak of the "environment", what we really mean is a relationship existing between nature and the society which lives in it. Nature cannot be regarded as something separate from ourselves or as a mere setting in which we live. We are part of nature, included in it and thus in constant interaction with it.'

- Thus, we may synthesise his various pronouncements and treatment of the ecological question (natural & human ecology or integral ecology) under four headings:
- The call to protect (environment and life) is integral and all-embracing.

- The care for creation and the respect for the grammar of nature are virtues in their own right.
- The need for education in ecological citizenship to underpin a moral conversion: an ecological conversion to care for what we cherish and revere.
- The recognition that binding regulations, policies and targets are necessary tools for addressing poverty and climate change, but they are unlikely to prove effective without moral conversion and a change of heart. Our efforts at combating, mitigating, or preventing climate change, global warming, poverty, and inhuman conditions require an integral approach to ecology. It cannot be limited to legislation, policies, or merely scientific, economic, or technical solutions. To succeed, whatever is done must be undergirded by an 'ecological conversion': a real conversion of mind, heart, and lifestyle, and in a new global solidarity.

These words of Pope Francis bring to mind Pope John Paul II's call for a 'moral structure' to the ecological question, and Pope Benedict XVI's advertence to the 'moral tenor of society' that is required to deal with the ecological question.

B) The Holy See and international organisations on water (oceans):
Equipped with this tradition of Church teaching about access to created goods as a right and a common good, even in the different phases of its development, the Church has and continues to engage in international meetings and discussions about such created goods as water, the sea, rivers, groundwater, and glaciers as well as sustainable development more generally (e.g. SDG 6).

Already in 1977, during the third UN Conference on the Law of the Sea (1973-1982), the Holy See Dicastery: Pontifical Commission: Iustitia et Pax, contributed a Working Paper on the universal purpose of created things.

Subsequently, and skipping for now a consideration of the UN MDGs and the SDGs, we may consider the participation of the Holy See in the World Water Forums:

World Water Forums

Between 2003 and 2012, the Holy See, through its Dicastery for Social issues, the Pontifical Council of Justice and Peace, participated in four World Water Forums, accompanying discussions with a call for the moral consideration of issues. Thus, at Kyoto (2003), the Holy See contributed to the discussion with a position paper, entitled 'Water, an essential element for life'. In this paper, the Holy See drew attention to how water fulfils a basic need for the three pillars of sustainable development, namely: the economy, society, and the environment. It described three water-related goods: 'economic good', 'social good', and 'environmental good'. Recognising the lack of these three goods for some people, and the lack of access to potable water in certain areas of the world, the paper raised the issue of access to water as a common good, a necessity of life, an expression or a requirement of a person's dignity and, therefore, a 'right'. Moreover, given its great value and importance the management of water against abuses is an absolute necessity.

At the Forum in Mexico City (2006), the Holy See built on its reflections at Kyoto, now signalling water as a fundamental good of God's creation, destined to serve the good or the well-being of every person. In this sense, the position paper presented at Mexico City identified water as a key factor for peace and security, and stated that created goods are the 'responsibility of all.' In this regard, the paper called for the promotion of a 'culture of water' that values and respects it, and desists from treating it as mere merchandise or a marketable commodity. Hence water must be managed with a keen sense of justice and responsibility.

The 5th Water forum in Istanbul (2009) was held against the background of several water challenges in the world. Accordingly, the Holy See position paper bore a sense of urgency. It added to a subtitle to its Kyoto paper: 'Water, an essential Element for Life: And now a matter of greater urgency.' This urgency originated in a persistent lack of access to potable water and hygienic services or sanitation, which makes the right to access to water elusive. Accordingly, the Holy See position paper insisted on a clearer legal formulation of people's right to water.

The World Water Council convoked the 6th World Water Forum in Marseille (2012) under the banner: 'Time for Solutions'. In view of, particularly, two issues: trans-boundary water management and green growth (food security), the Holy See's intervention focused on the role of water for peace and conflict. Motivated by the OECD's call for worldwide water reform, the Holy See delegation highlighted the need to clearly define and uphold the water rights of people and the centrality of the human person, his/her access to created goods, and the primacy of his/her dignity in the management and legislation of created goods, like water. The World Water Council and the Holy See's dialogue continues.

The oceans and the seas

In June 2017, the UN held a High-Level Conference on the 'Oceans and SDG 14', to which I was privileged to lead a delegation of the Holy See. The thrust of the conference was to 'conserve and sustainably use the oceans and marine resources for a sustainable development'. It featured several partnership dialogues, two of which addressed the issues of 'minimizing and addressing ocean acidification, and increasing economic benefits (of the Blue economy) to Small Island developing States and least developed countries, providing access for small scale artisanal fishers to marine resources and markets.'

For the Holy See, taking care of our environment, a gift entrusted to our responsible stewardship, is a moral imperative. Among the many considerations that flow from this fundamental principle are intergenerational solidarity and a focus not merely on rights, but also on responsibilities. Pope Francis has repeatedly affirmed that intergenerational solidarity is not optional, but a basic question of justice, since the world we have received also belongs to those who will follow us. This is reason enough to consider the impact of anthropogenic climate change on the oceans. The oceans and seas should immediately be considered a benefit. They must be viewed as a gift for future generations, too. Moreover, we must spare them from paying the extremely high price of the deterioration of our oceans, seas, and marine resources.

In July 2017, the ambassadors of Monaco, France, and the Netherlands accredited to the Holy See organised a conference with the Dicastery for Promoting Integral Human Development of the Holy See at the Pontifical University of the Holy Cross (Rome), on 'Care for Oceans: The Oceans, Caring for a Common Heritage'. The United Nations was represented at that conference by Mr. Peter Thompson, President of the UN General Assembly; the focus of the event was the impact of climate change on the oceans and seas. The Dicastery's contribution was a reminder that the oceans and their underwater treasures and wonders constitute a 'heritage' to humanity. 'Heritage' makes us recognise that we have inherited these created goods and have a duty to pass them on. It calls for a keen sense of responsibility and of intergenerational solidarity as caretakers.

Many other states, such as Indonesia, the Netherlands, Chile, the United States and Fiji, have held international conferences on the ocean. The decision for Fiji to jointly host the COP23 with Germany, in Bonn (2017), was a signal that we must address the 'cry for help' from Pacific Island States slowly sinking under rising sea levels as a result of climate change. At both COP23 and COP 24, held in Katowice, Poland, the Holy See appealed for a show of 'solidarity' for peoples dealing with the effects of climate change.

In October 2017, the European Union held a conference in Malta for 'a safe, secure, clean and a healthy ocean.' In the background were the challenges of

African migrants, compelled by climate change conditions to risk crossing the Mediterranean Sea. And a UN working group at the last meeting of the International Maritime Organization in London (26-27 November 2019) eagerly committed to preparing "a new legally binding instrument to expand the Law of the Sea to regulate the conservation and sustainable use of marine biodiversity in areas beyond national jurisdiction in these times of climate change.' The Holy See was there to offer the help of its special apostolate for seafarers and fishers, called 'The Apostleship of the Sea'.

The Stockholm SIWI World Water Week

Since 2016, the SIWI World Water Week has offered the Holy See (the Dicastery for Promoting Integral Human Development) an annual occasion to engage the world of science, economics, politics, as well as professionals on various water-related issues, such as water governance or management, but also water-related challenges, such as the water-climate-economics and poverty-health-sanitation chains.

Thus, the 2016 World Water Week was inspired, initially, by the Encyclical Letter of Pope Francis, *Laudato Si'* par.14: 'I urgently appeal, then for a new dialogue about how we are shaping the future or our planet. We need a conversation which includes everyone, since the environmental challenge we are undergoing, and its human roots, concern and affect us all.' On this basis, the SIWI meeting on 'Faith, Water and Sustainable Development' welcomed input from religious leaders on how their different faiths promote water management for sustainable development in an era of climate change. The Holy See delegation contributed the concept of 'integral development' as a programme of inclusive development that involves the growth of the human family and the well-being of its environment. Well-pursued 'integral development' necessarily requires the adoption of measures of mitigation and adaptation to reduce the impact of climate change. Most importantly, the contribution of water to integral and sustainable development, in situations when the variations of the water cycle are causing climate change, requires the development, by all, of a serious water culture, a keen education in ecological citizenship for an ecological conversion that enshrines a deep respect for created goods, the miracle of creation, and a profound sense of stewardship or 'caretaking' of God's gift to humanity.

Reflection by Hein Pieper

VICE-PRESIDENT OF THE DUTCH UNION OF WATER COUNCILS

The Netherlands has a long tradition of water management: some say that God did not create our country, but we did it ourselves. The history of the Dutch water authorities begins around the twelfth century. Previously, mainly abbeys and monasteries were responsible for water management. The twelfth century was a special period for our region – the Netherlands did not yet exist as a nation. It flourished as part of a wider Renaissance that continues to exert its influence today. Poverty was less prevalent at that time than in the so-called Golden Age. Latin schools provided education, abbeys and monasteries offered healthcare and good management of the first public goods. It was an era of common ground and common responsibilities. The values in force at the time were converted into a mentality that resulted in a prosperous country. The water authorities were created in this context –a combination of faith, values, knowledge, and resources.

In our highly technical world that relies on instrumental rationality, procedures, protocols, business plans, models, and algorithms are in the forefront. Despite the major developments that have been achieved, we are confronted with the negative sides of this prosperity. Climate change is one of them and increasing water problems are a major concern. Answers to these major problems are slow in coming. Goals such as the Millennium Goals or the Kyoto targets are not achieved. The solution is always sought in the future, data is pushed back, making it seem that the problems are still soluble. 2000 became 2020 and now we are talking about 2050 or 2100. For me, the problem lies in the way in which answers are sought. It is always the same kind of answer, based on instrumental rationality. Automation becomes robotics. IT becomes Big Data. Models become new models. Climate change primarily requires a conversion of the heart. We need a societal transformation. Our way of life must change fundamentally. Knowledge can only serve such a process.

In recent years, the Dutch water authorities have increasingly received requests to host foreign delegations or to assist with problems concerning water management abroad. All of these issues have been or are affected by climate change. Hence I launched the Blue Deal in 2018, a project that will run until 2030. The Blue Deal aims to improve the water availability, water quality, and flood protection for 20 million people in 40 catchment areas all over the world by 2030. It is a project of the Dutch Water Authorities and local or regional water authorities abroad, supported by two Dutch Ministries, the Ministry of Foreign Affairs and the Ministry of Infrastructure and Water Management.

The water authorities would like to share their knowledge, but acknowledge that it goes beyond knowledge sharing. In this regard, the symposium organised by the Vrije Universiteit 'Water in Times of Climate Change: A Values-driven Dialogue' together with the Patriarch of the Orthodox Church was a welcome addition to the work of the Blue Deal. This symposium went beyond the traditional approach to water management and climate change in an innovative way. A promising and surprising interaction was initiated between the dimensions of technical solutions community values, religion, economy, ecology, and politics, with a focus on the urban areas of Cape Town, Jakarta, and Amsterdam. This approach helps to explore new perspectives in the world of water management. We hope that this collaboration can grow into concrete long-term projects. We are currently investigating whether we can start a joint project between the cities of Amsterdam, Cape Town, and Jakarta. This could become a wonderful example. We look forward to the results, because exchanging knowledge and experience between these community-based water-related projects and the Blue Deal South Africa Partnership is mutually beneficial. Some of our projects are based on the same principles of a bottom-up approach with social and environmental community-based values. By sharing the knowledge and experiences of our community-based activities and using the 'best practices' applicable, we hope to build a new, long-lasting relationship and find answers to the major biases of our technocratic times. The Blue Deal projects are taking place all over the world and are still growing. This concrete project can be an example for many others. I sincerely hope that we can achieve the social transformation needed to combat climate change. If we stick to the prevailing technocratic approach, we will fail. Rather, a values approach is needed, in which churches and religions are indispensable. In this regard, *Laudato Si'* is an inspiring guide.

IJVEER 61
GVB

5.

Water Sensitive Cities

This chapter addresses water in times of climate change from the vantage point of three major cities: Jakarta, Amsterdam, and Cape Town. In the previous chapters, we have seen that water is a key factor in some of the most compelling challenges people have to meet worldwide. Rising sea levels, drought and desertification, shortage of drinking water, and sanitation shape and form our struggles with water that will be crucial to the sustainability and viability of the earth. Nevertheless, a politics of hope does not live in abstractions. Hope can become visible in relations and situations that are under real threat. Therefore, in this chapter we focus on three cities that are threatened by water in times of climate change. The three key themes identified in the introduction are used to categorise the cities:

1. Cleansing water: Jakarta, fastest sinking city
2. Threatening water: Amsterdam, city two metres below sea level
3. Life-giving water: Cape Town, first city to run out of water

Each of the three cities provides relevant insights on how to become a water sensitive city. Participation of city officials, scholars, business, artists, and religious communities serves to develop common language and to strengthen local and global partnerships.

DA VINCI
DA VINCI TOWER

KELUAR
Angke
Pejagalan
Glodok
500 m
Grogol
Tomang
Jkt-Tangerang

Jakarta

Government: Elisabeth Tarigan, Water Resources Department (DKI)
Science: Inke Prima Diyarni, Development Planning Agency (DKI)
Economics: Feli Napraieti, Human Settlement, Spatial Planning, Land Authority Department (DKI)
Religion: Haryani Saptaningtyas, Percik Institute Indonesia

Bengawan Solo

Air mengalir sampai jauh,
Akhirnya ke laut

Water flows far away,
Finally to the sea

JAVANESE FOLK SONG

Introduction to Jakarta

With a coastline of c. 81,000 km and more than 17,500 islands, Indonesia is extremely vulnerable to coastal inundation. Jakarta, Indonesia's largest city, is located in a lowland area with a relatively flat topography in the delta of several rivers, the main one being the Ciliwung River. Due to its naturally flood-prone location, Jakarta has a long history of both coastal and riverine flooding.

- Capital of the Republic of Indonesia
- Population: 14.2 Million (day), 10.3 Million (night)
- Area Size: 662,33 km² (land) and 6,977.5 km² (counting five administrative cities and one regency)

Jakarta in times of climate change

Indonesia is one of the most vulnerable countries to climate change due to its geographical, physical, and social-economic situations. There are many initiatives to understand and deal with the impacts in the country. The national government has issued key guiding policies for climate change. International agencies work together with local stakeholders to strengthen capacity in policy formulations and implementing actions to build community resilience. Universities conduct research on climate change related at different scales. Cities and local governments implement innovations in adapting to the impacts of climate change and transitioning towards a green economy. The city of Jakarta is working a reduction plan to deal with greenhouse gasses and pollution. With this plan, the city government has set itself a goal of reducing GHG emissions by 30 per cent by 2030.

From a government perspective: The Jakarta Provincial Government understands the importance of water and is committed in its Provincial Development Plan to achieving the improvement target with respect to water management, focused in particular on flood disaster, clean water supply, and wastewater management. The Master Plan for Wastewater Management in DKI Jakarta (March 2012) forms the

basis for improvement of water quality in Jakarta. The vision laid out in this Master Plan is to improve the current river water quality to the level that river water can be used as a source for the water supply system in DKI Jakarta by 2050.

From a scientific perspective: Dr Heri Andreas from the Faculty of Earth Sciences and Technology, Bandung Institute of Technology has developed a model of Jakarta's land subsidence through the years (see below). By 2050, the capital of Indonesia will be entirely submerged and 95 per cent of north Jakarta will be underwater, directly impacting 1.8 million of the city's ten million people. Scientific observations and climate model results indicate that human activities are now the primary cause of most of the ongoing increase in the earth's globally averaged surface temperature. That is why it is important to reduce land subsidence by reducing groundwater abstraction through provision of alternative water sources such as recycled water.

From an economic perspective: Climate change impacts will affect the number of costs necessary for adaptation and mitigation. In Jakarta, the elements of adaptation costs are all budgeted to support climate change adaptation's policies, such as vertical drainage construction, rainwater harvesting movements, installation of Giant Sea Walls, river dam reservoirs revitalisation, and providing alternative water sources through recycling wastewater. Meanwhile, the mitigation cost element is a budget that must be prepared and used in response to the impact of climate change, for example, health expenses or the cost of rebuilding infrastructure in areas affected by flooding. Thus, it is expected that there will be an increase in the quality of urban planning, risk management, citizen awareness, and public engagement that will reduce the economic impact of climate change in Jakarta.

From a religious perspective: Religions in Indonesia not only provide social identity for many people, but also provide ethics that may be used to create behavioural change. Religion can be a hindrance in applying technological innovation to, for example, the reuse of ablution water, unless it is followed by religious interpretations and the need for a new awareness regarding environmental problems. Therefore, solving water problems requires a dialogue between religious understanding, technological innovation, and medical perceptions.

Amsterdam

Government: Kees van der Lugt, World Waternet
Science: Esseline Schieven, City of Amsterdam
Economics: Marieke Abcouwer, ABN AMRO
Religion: Rabbi Awraham Soetendorp

"Amsterdam die grote stad, die is gebouwd op palen. En als die stad eens ommeviel, wie zou dat dan betalen?"

"Amsterdam, that big city built on top of long wooden posts. And if that city toppled over someday, who would pay the cost?"

JOHANNES VAN VLOTEN, TAKEN FROM *BAKER- EN KINDERRIJMEN* (1894)

Introduction to Amsterdam

Once directly connected to the North Sea and the world, Amsterdam was the most important harbour in the world. Since then, the city has become more protected from the sea thanks to dams and water works. But the famous structure of finely meshed canals remains a cultural heritage from the days the Dutch sailed the world seas.

- Capital of the Kingdom of the Netherlands
- Population: 1,148,972
- Size: 219.5 km^2

Amsterdam in times of climate change

Through the centuries, the inhabitants of the Netherlands have struggled with the constantly changing situations of rivers and the sea. The low-lying country

has always been under threat of the powerful sea. Through collaborative water management (the so-called Polder model) the Dutch kept their feet dry with an occasional water disaster that brought new focus. Consequently, the country has played a central role in the development of innovations related to water management. One can refer, for example, to the disastrous floods in the southern part of the Netherlands in 1953. From this catastrophe came the impressive and famous Delta Works. Given the small size of the country, the broader context has always played a major role in the development of the Netherlands. Today, nine million people live below sea level in the Netherlands, a number increasing through densification and sea-level rise. Roughly 65 per cent of the Dutch gross national product (GNP) is earned in this part of the country. As a follow-up to the Paris Climate Agreement, the Netherlands has passed an ambitious climate act with a target of 49 per cent greenhouse gas emission reduction by 2030 compared to 1990 levels and a 95 per cent cut by 2050. Key questions remain, however, about how to adapt to the predicted sea-level rise, greater fluctuations in river discharge, heavy rainfalls, and drought, and how to shape the required transition to a greener economy and society.

From a government perspective: Politics operates on a short-term basis. From a classic polder mechanism focused on consensus and incremental changes, we see a higher dynamic in the debate through populism and polarisation. Complex problems are often questioned and comfortably directed towards a future with a silent hope that science and economics will tackle the problems. The dogma of economic sanctions is paralysing.

From a scientific perspective: Science addresses the problems of climate change and focuses less on solutions. The field is under credibility stress. Is there a truth in science or is it just an alternative? Are the presented facts absolute or is the focus too much on the uncertainties in the super-complex field of climate studies? Can we cope with some of the predictions on our future? What does substantial or extreme sea level rise mean for living in river deltas?

From an economic perspective: Economics is driven by the classic endless promise of growth and individual possession. Unfair pricing of products and

services took an enormous mortgage out on the future. Who should pay the interest, if, indeed, we can ever pay it back? Climate stress on investments in industry, agriculture, or real estate will shift balances towards more solid investment portfolios. In the future, the assets will not be frozen but drowned. Money travels fast, is there a climate crash in a nearby future?

From a religious perspective: In early times, religion gave societies hope and direction when uncertainty was a certainty. It became less dominant in societies where safety, wealth, and certainties could be created through the triumvirate of science, politics, and economics. Religion comforted people and society in times of stress through grief, mourning, acceptance, and hope. Is religion helping us in these times of new uncertainties and able to bring new hopeful perspectives after the grief of change? Hope is just beyond our reach. The only thing we have to do is to respond to its call. Each generation, of both believers and non-believers, is challenged to carry together that precious torch of hope into a radically uncertain future.

we tame the water

by Sjaan Flikweert

there is an ocean
in these empty sheets of paper
where we write stories

an ocean of ink in fingers
there is an ocean in the birth of ideas
feelings of hope in not understanding
the howling waves

we go on vacation to the horizon
listen to the boasting waves
daring them: come and wrestle with our virtues
they will spit our bodies on the beach
and skin our egos to the bone

here we take vacations from
coffee breaks and the water bill,
torn shoelaces and the gym

we are nothing, nobody
set side by side the ocean

while wrestling the waves
we inhale a gasp of salt water
later the sun will wring it out of us
the leftovers will fall like tears

on the way back we climb dunes
soon all will be, as it always has been
it matters what we do between the levees

we made this land
we tame the water, here we call the shots

we tame time in schedules and checklists
we tame plants and small talks
we regulate the humidity and create hydrologic cycles
we mourn our pet fish
it matters what we do
we tame water

our dogs listen, because we feed them
this is also why our cats return home
water knows no death no life,
gives birth to crosscuts
it does not care
it does not need us

there is an ocean
in these empty sheets of paper
where we write stories

an ocean of ink in fingers
there is an ocean in the birth of ideas
feelings of hope in not understanding
the howling waves

we tame water, even though she does not need us
we are the ocean, she is no human
she spits and no mountain can stop her

let's descend and see her for who she really
is the untameable

LAW
EZENDLELA

Cape Town

Government: Michael Webster, Director Water and Waste City of Cape Town
Science: Kevin Winter, University of Cape Town
Economics: Charon Marais, Stellenbosch University
Religion: Thabo Makgoba, Archbishop of Cape Town

I want to wake up to the Majestic Table Mountain
Always standing, Always there, Never failing
Its beauty astounding
As it watches over Cape Town...
I want to go home to gaardtjies shouting
'hullo, hullo, Mowbray...Cape Town' with their gapped smiles.
Never mind the way they weave between the traffic.
I want to go home to a place where people say
'just now' for later, 'kanala', 'shukran' and
'slamat on your birthday'...
I want to go home...

FAIEZA MX[6]

Introduction to Cape Town

If three pyramids are the symbol of Africa's far north, then a flat-topped mountain is the symbol of its far south. The city of Cape Town is a large urban area with a high population density, an intense movement of people, goods and services, extensive development and multiple business districts and industrial areas.

6 Available from https://www.sapeople.com/2015/04/18/i-want-to-go-home-to-south-africa/ [Accessed 5 March 2021]

- Legislative capital of the Republic of South Africa
- Population: 4,618,000
- Size: 400.3 km

Cape Town in times of climate change

In January 2018, news of the water crisis in Cape Town ricocheted around the world. Officials in Cape Town announced that the city was three months away from 'Day Zero' – the day when the taps would run dry. Brought on by three years of below average rainfall, the city faced the title of 'the first city to run out of water'. In the end, Day Zero did not arrive due to a sharp reduction in water demand and a return of the rainfall. Nevertheless, there is a growing consensus that the city is vulnerable to climate change.

From a government perspective: Despite the distractions of finger pointing between national, provincial, and local government as to who was responsible for the water shortages, the city of Cape Town acted quickly and implemented a highly successful water demand management strategy. Inspired to further action, the city launched its water strategy 'Our Shared Water Future: Cape Town's Water Strategy' on 19 February 2020. This document provides a roadmap towards Cape Town as a water sensitive city in 2040, in which there will be sufficient water for all, and the city will be more resilient to climate and other shocks.

From a scientific perspective: The answers to water science challenges are no longer limited to the world's engineers. Rather, a multidisciplinary approach (health sciences, economics, law, humanities together with sciences and engineering) is needed to solve the water management challenges in South Africa.

From an economic perspective: The water crisis in 2018 had a significant impact on the region's economy. Both agriculture and tourism, key economic sectors for the region, suffered significantly and many jobs were lost. Other specialised commercial businesses were impacted by the drought, with some actually benefiting and many others suffering (depending on their line of work). This 'new normal' meant that moving forward, in order to ensure sustainability, all businesses active in the city needed to 'water-proof' their businesses in not relying on municipal water provision.

From a religious perspective: As tensions rose in 2018, faith groups began working across spiritual divides to offer their congregants hope and a way forward. Beyond individual faith groups' efforts, several ecumenical and interfaith initiatives emerged to address the arising challenges. Archbishop Thabo Makgoba of the Anglican Church challenged all citizens as follows: 'When the interfaith voice opposed apartheid as a movement, it worked. This is a struggle and a crisis. We need to be good stewards. It is a beautiful opportunity for South Africans to come together. *And therein lies our hope.*'

From an arts perspective:

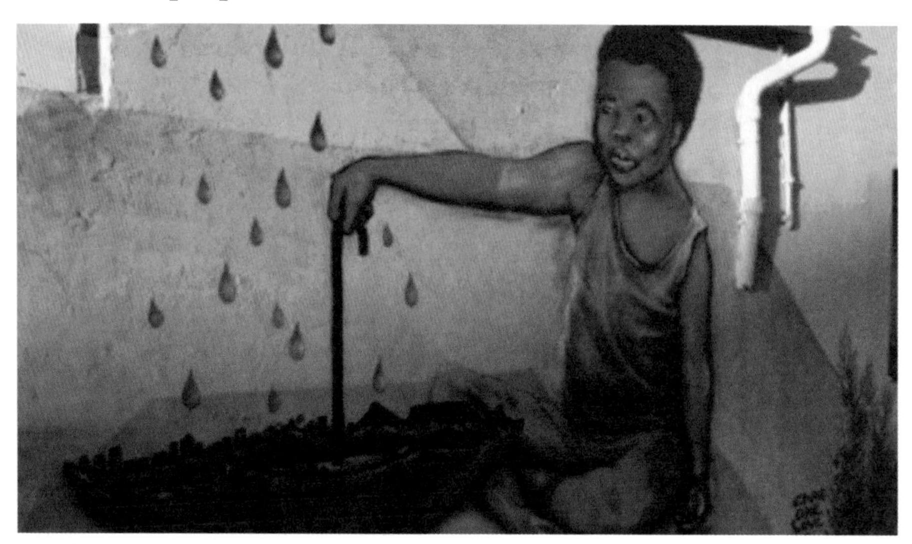

(graffiti artist: CareOneLove)

Reflection by Thabo Makgoba

DR THABO MAKGOBA, ARCHBISHOP OF CAPE TOWN

In 2018, we were told that most taps across Cape Town, a city of 4 million people, would be turned off on April 22. 'Day Zero' would arrive when the water in the dams supplying Cape Town reached less than 13%. Schools and businesses would have to shut, people would have to queue to fetch their water, and the police were to be mobilised to quell potential fights over access to water. After three years of the driest winters on record and a one-in-300-year drought, the dams were nearly empty. Cape Town rallied as never before and we reduced our water demand by a massive 50 per cent. Week by week we pushed back Day Zero – until the winter rains came again.

How did the faith communities respond? Initially, we were on our knees praying for a miracle. But then we realised that we must pray for something deeper – we must realise again that water is sacred, a gift from god. We had reduced water to a commodity. We had lost the sense of sacredness of water, seeing it as 'something that comes out of a tap'. As Christians, we become members of the family of God through the sacred waters of baptism, water is our primal element. Similarly, our Muslim brothers and sisters wash their hands before they pray, five times a day. Water flows through the pages of the Bible, the Koran, and many sacred scriptures. At the start of creation we read that 'the spirit of god was hovering over the waters.'. At the end of the Bible, the Book of Revelation tells us that 'the waters of life flow from the throne of god.' Water rushes, gushes, and pours through the pages of the Bible in 722 verses! We must recognise again that water is sacred and commit to protecting it. If God has placed the care of this planet into our hands, then relying on prayer alone and doing nothing to safeguard this precious gift is theologically irresponsible. In 2018, churches started committing to water harvesting and use of greywater, checking for leaks as they called for rain in their prayers.

Secondly, we awoke to the reality of water justice – for many people in our city it is Day Zero every day of their lives because they must carry water from communal taps to their homes in informal settlements. It is said that you do not realise the value of water until you have carried it. Cape Town is one of the most unequal cities in the world. On one side are homes with more bathrooms than people, with big swimming pools and vast lawns for a couple of children to play on. On the other side, nearly 20 per cent of Cape Town comprises informal homes where five or more families share one communal toilet and tap and, more often than not, the toilets are out of order. Girls are afraid to use the communal toilets at night for fear of being attacked and raped, and during the day children often play in the filthy water seeping from the poorly serviced toilets and polluted water running from homes. Clean drinking water and sanitation is the most fundamental human right. It is central to the well-being of all people on the planet and the lack of access to clean, fresh water is one of the most serious threats to human health.

The book of the prophet Amos poses a strong challenge: 'let justice flow down like rivers, and righteousness like a never-ending stream' (Amos 5:24). So, how can we, as the Church, respond to the challenges of water? We must protect our rivers and aquifers: in South Africa, half of our surface water comes from only ten per cent of the land – these 'catchment' areas must be protected as a priority. Cape Town is the only city to contain a catchment area – the Table Mountain water source area. The city's water demand has grown so it now draws most of its water from other mountainous areas outside the boundaries of the city. What a wonderful picture of 'water coming from the throne of god' – water flowing from the beautiful mountains that surround us. Water that must be treasured and protected.

The challenge is not just that the rains do not fall – many of the threats to water come from companies who pollute rivers with industrial pollution. In South Africa, we suffer greatly from acid mine drainage affecting our water systems. The shareholders of mining companies make a profit, but the local communities are left with water degradation. Large corporate farms are also responsible, as the run-off from artificial fertilisers and pesticides pollutes the rivers. We are holding 'courageous conversations' with the mining companies to challenge them on some of these environmental issues.

Churches are adopting local rivers as part of their spiritual journey – recognising them as their own 'River Jordan' – holding regular clean ups, putting in litter traps, planting trees, and taking shared responsibility. To care for our rivers we must become passionate campaigners against the scourge of plastic waste. At our last synod, the Anglican Church of Southern Africa passed a motion calling for a ban on single-use plastic.

If we want to fight for water justice we must also fight for climate justice. Studies tell us that the results will be devastating if we do not act fast in the next 11 years. The rise in temperature threatens farming, dam levels, and our natural environment: plants, birds, and vulnerable amphibian species. Our food supply, water security, and economy will be under threat. We may not be around to see the worst impacts of climate change, but our children will be. In the end, the harsh uncomfortable truth is, we do not inherit the earth from our ancestors, we are stealing it from our children. The Bible tells us in Romans 8:22 that 'creation is groaning as in childbirth.' This is an image of great suffering, but ultimately it is a vision of hope – the birth waters will break, and we will see new life come forth, clean water flowing from the throne of God and bringing healing to the nations. 'Creation is standing on tiptoe waiting for the sons and daughters of god to be revealed.' We have delayed too long; it is time to respond.

The most important message is this – if you want to do something, then change your lifestyle, influence your congregations and communities, and challenge your politicians and let us make fighting climate change the highest priority on all of our agendas.

Let me end with a challenge to us all. We have lived our lives by the assumption that what was good for us would be good for the world. We were wrong. We must change our lives so that it will be possible to live by the contrary assumption, that what is good for the world will be good for us. This requires that we make the effort to know the world and learn what is good for it. It means living for the good of the community, for all of us, not living for our own individual ends. It means living the we and not the me.

In sum, let us end the indifference.
Let us end our indifference to the importance of water and its centrality in our faith and our lives.
The opposite of love is not hate – it is indifference.
The opposite of faith is not heresy – it is indifference.
The opposite of life is not death – it is indifference.
The opposite of hope is not despair, but indifference.
May we become agents of hope.

Available in 3 Colors
7

6.

A Covenant of Hope

During the closing session of the symposium 'Water in Times of Climate Change: A Values-driven Dialogue' parties signed a covenant of hope. A covenant is one of the coordination mechanisms or rituals, highlighted by religious traditions, to foster hope, trust, and solidarity in the midst of a complex reality that includes conflicting interests, dead ends, pure self-interest, and fear. Hope neither rejects, nor surrenders to this reality. A covenant of hope enters into being when two or more parties voluntarily promise to take responsibility for a shared future. The covenant values the plurality among participants. Each participant becomes part of the covenant on his or her own terms. As a consequence, to be part of the same covenant does not mean that everybody agrees with one another. A covenant is an argumentative association. The covenant does not seek the affirmation of one position, rather it stimulates opposition to open the identities of those involved in order to create a new and common identity.

The present covenant of hope builds on a series of workshops that were held during the symposium. In these workshops, different parties came together to discuss topics such as *Laudato Si'*-proof financing, working together with water, stakeholder engagement, how the sacred scriptures can help us in times of climate change, and what it means to build a politics of hope amidst threats. The workshops were considered to be a safe space to build relations and develop

dialogue between cities or between the interlocking dimensions of science, politics, economics, and religion. Each workshop resulted in one or two recommendations. In the conclusion of this book we describe various initiatives that have come out of these recommendations. In this chapter, we present the full text of the covenant followed by reflections from Patriarch Bartholomew, Gerhard van den Top, Jos Douma, and ABN AMRO bank.

apg
accenture

Covenant of hope

Water: source of life, symbol of purity. But also a threatening force of nature that humans have to struggle with. Life-giving friend, life-taking foe. Since time immemorial and across the globe this ambiguous relationship with water has resonated in religious narratives and technological innovations alike. Today, it also resonates in several of the Sustainable Development Goals, the umbrella for a programme addressing the challenges of our times.

Securing our existence and the future of our children has become more than navigating ambiguity. Water in times of climate change has become a radical uncertainty, key to the most compelling challenges of our societies. Rising sea levels, drought and desertification, shortage of drinking water and sanitation, shapes and forms our struggles with water that will be crucial to the sustainability and viability of the earth.

We can respond in various ways to this radical uncertainty and ambiguous complexity. Reckless denial ignores all the warning signs and postpones all action so that the next generation will suffer the consequences. Helpless despair turns to an overwhelming dread of the consequences, which, in turn, saps our power and courage to act. Thoughtless self-confidence believes that our technological ingenuity will suffice so that we risk overlooking moral dimensions and yet unseen complexities.

Our response is a fearless hope that acknowledges uncertainty and complexity. Hope balances the imperfections and failings of the present with the promises and possibilities of the future. Hope builds the bridge between the 'what is' of reality and the 'what if' of our visions. Hope is the opposite of denial, of despair, and of the self-confidence that so easily turns into a new escapism. Hope is the engaged and engaging response of the people of today to the calling from the future.

This covenant of hope invites us to respond to that calling. It brings together all those of good will, ready to share our insights, visions, resources, and capabilities. The covenant respects the dignity of our differences and the responsibility for joint action. The covenant seeks to bridge our practical, technological, legal, economical, and spiritual understandings of our predicament. Together, we will take the small steps needed today to reach our rich vision of living sustainably on this earth, living with water as our dangerous friend.

Reflection by Ecumenical Patriarch Bartholomew

ECUMENICAL PATRIARCH BARTHOLOMEW

At the end of the Orthodox service for the 'Blessing of the Waters', which we observe on the first day of each month, every person in the congregation is sprinkled with water and takes home a small bottle to bless their house and family. As we conclude, what will each of us take home to touch our family and friends, while at the same time to transform our work and society?

One of the most powerful aspects of water is that it flows. And when it flows continuously, it remains purifying and life-giving. There would be no reason or meaning to gathering if we did not practically reach out to all those with whom we come into contact and on whom we could potentially have some impact.

As we have emphasised, water is the inviolable and non-negotiable right of every human being and every living thing. We must publicly declare our conviction that the water crisis is both moral and urgent; that whether withholding water from our brothers and sisters or exploiting water through abuse or waste, we are harming not only ourselves but future generations. We are both responsible for the sustainability of our water and accountable for the survival of our children.

Whether living in cities or continents – in Amsterdam, Africa, or Asia – our aspiration is to preserve and share clean water for every human being. Without water, there is despair; but with water, there is hope. We all need water in order to survive and thrive. In fact, the body can endure longer without food than it can without water. We are called to remember and to realise that the cycle of water and the cycle of life are one and the same.

This means that our connection to water should reflect our connection to other human beings. Our treatment of water should echo our respect for other people. Moreover, our relationship with the earth's water is not a commercial contract, but a spiritual covenant with all living creatures and the whole of creation. We must expand our concept of God's creation and compassion as relating exclusively to human beings. God's covenant is cosmic; and the philosophical shift that we are called to make is cosmological. The level of our greediness and

wastefulness is an ethical challenge that was unimaginable to previous cultures. Nature will be fruitful and plentiful when human beings respect the covenant between God and creation.

Such was the promise that God offered to Noah and his children in Genesis (9:12-17): God said to Noah, 'This is the sign of the covenant I have established between me and you and every living creature [...] a sign and a covenant for all generations to come [...] an everlasting covenant between God and all living creatures of every kind on the earth.' This, however, is far more than mere stewardship and sustainability. It is a pledge – a spiritual contract and societal assurance – that all of us owe to our own children for all generations to come.

Take this with you to your homes and communities, your places of work and interaction, as well as your cities and countries. May God bless you abundantly. And may God's living water and spring of grace refresh, guide, and inspire you all the days of your life.

Reflection by Gerhard van den Top

CHAIRMAN, WATERNET AMSTERDAM REGIONAL WATER AUTHORITY

When you enter the central hall of our Water Authority headquarters, to your left you will see a silver chalice dating back to 1717. This 'Chalice of Cooperation' (in Dutch: *Hensbeker*) was used by the boards of the water authorities as a symbolic tool to confirm agreements that were meant for the greater good. It symbolised cooperation and the significance of new agreements and decisions in the area of Amsterdam.

The Chalice of Cooperation is part of a broader Dutch governance tradition, going back to the year 1000, when growing settlements on the sea-river interface of 'the lowlands' began to realise they needed to take responsibility together in the management of shared water challenges and ambitions. Some argue that the typical Dutch way of dealing with major societal issues in a diverse society through slow processes of trust and consensus building, called the 'Polder model', can be traced back to the way we have learned to deal with water challenges.

Although we still use the chalice on special occasions, today it is mostly preserved in a cabinet or a museum for its historic value. But we cannot afford to preserve the spirit of the chalice in a cabinet. When I look around I see, on the one hand, a widening and deepening polarisation on societal issues – agriculture, water, climate change – and, on the other, an increasing urgency to come together around these issues. There is a great need for initiatives that can stimulate a spirit of cooperation and inclusiveness, enabling all stakeholders to share their wants and needs in dealing with these common challenges. We need hope over fear, and steps forward together over paralysis.

To contribute to this spirit, we re-introduced the symbolism of the historic Chalice of Cooperation during the closing ceremony of the Amsterdam Water Week 2019. The old crests of the historic water districts that were engraved in the original silver chalice were replaced by an image of the world. This image of our globe represents cooperation on water issues between cities and utilities worldwide in times of climate change, and between the 'worlds' of science, economy, politics, and religion. The image was printed on a reusable '*Dopper*' bottle, to inspire stakeholders to also choose reusable over single-use water bottles. The old-new chalice was used to underline the importance of achieving breakthroughs by new coalitions, 'Amsterdam Agreements', for the urgent existing water challenges.

Thus, the 2019 edition of the Amsterdam International Water Week became an example of bringing the language of ritual and meaning into a world often dominated by the language of natural science and technology. So that further cooperation between natural science, economy, politics, and old-new rituals of meaning may guide us to more water sensitive cities.

Reflection by Jos Douma

DUTCH SPECIAL ENVOY ON RELIGION AND BELIEF

It is now Easter 2020 as I reflect on the two symposia, 'Water in Times of Climate Change' in Amsterdam and 'Religion, Security and Peace: Religion and Belief in Contemporary Societies' in The Hague. Generally, such a 100+ days of reflection is simply good practice to evaluate and prioritise. But now a new paradigm – or at least a new perspective – imposes an even sharper focus.

The Covenant of Hope could still state that 'the issues that our world faces are in some ways hardly new.' But, wisely, it also noted that, 'our present situation is [...] quite unprecedented. [...] Never before has humanity been in a position to destroy so much of the planet environmentally. This predicament presents us with totally new circumstances, which demand of us a radical commitment to reconciliation and peace. The threat to the fabric of human life and the survival of the natural environment make this the overarching priority over all others.'

As much as in recent reflections on the COVID-19 crisis, human responsibility was central in the contributions to our seminars by Patriarch Bartholomew and Cardinal Turkson, also in those of J.P. Balkenende and others, both in Amsterdam and The Hague. The Covenant states (on climate and water issues) that 'we risk overlooking moral dimensions and yet unseen complexities.' And now, even more than ever, we should affirm that 'our response is a fearless hope that acknowledges uncertainty and complexity.'

When it comes to solving pressing issues and realising SDGs, I refer to the Ecumenical Patriarch, who ascribes to faith and religious groups a 'unique vantage point [...] to bring about a more "human" globalisation.' And, he continues: 'as well as advocating for peace and solidarity and against racism and discrimination, while at the same time championing religious tolerance and human dignity.'

To me, these are four major challenges to be faced post-seminars and post-COVID – 'with fearless hope.'

Reflection by Richard Kooloos

HEAD OF SUSTAINABLE BANKING, ABN AMRO

Water – a source of life and destruction – has sustained life on earth for centuries, often in extreme forms. The financial specialists at ABN AMRO MeesPierson Institutions & Charities know better than anyone that religious institutions and charities have long had a strong bond with this theme. Over the years, our clients have challenged us – and still do – to take our responsibility as stewards of the environment and to create forward-thinking sustainable solutions. At the same time, we want to share our network and expertise to stimulate the exchange of vision and knowledge. The water symposium gave us the opportunity to help build bridges between civil society organisations, but also between businesses, science, governments, and the younger generation.

In their own words: Rian Vens-Hagting, Director of Religious Institutions at ABN AMRO MeesPierson Institutions & Charities and Richard Kooloos, Director of Sustainable Banking at ABN AMRO.

Seeking connection

Throughout its history, ABN AMRO has worked with institutions and companies within the water sector – including a large number of religious organisations and charities – on promoting the role and significance of water in the world.

Climate change presents us with new challenges today, not only because of, but also with regard to, the role of water. There is an increasing awareness in the world that something needs to be done and that has opened the floodgates to a stream of initiatives. However, there has yet to be an all-of-the-above approach that would result in truly sustainable solutions. And when parties find each other, cooperation is often difficult, or even impossible. Why is that? It is often because they do not understand each other's language, do not know where to find each other, or sometimes it is because they do not trust each other.

The central purpose of *Laudato Si'* is summarised in a single question: 'What kind of world do we want to leave to those who come after us, to children who are now growing up?' The encyclical establishes the connection between man, ethics, and nature based on an integral ecology, the philosophy that the Green Patriarch has been actively propagating for 30 years. This year marks the fifth anniversary of ***Laudato Si'***, making this an appropriate time to reflect not only on how we have heeded the Pope's words, but also to issue a call to action to the financial sector to fight inequality and injustice with all possible means and contribute to a fairer society where possible. Embracing this important philosophy is an essential starting point for Rian Vens-Hagting in her search for a common language:

You have to tackle the problem and look for solutions together. Whether you are a church, dredging company, bank, charity or scientist, it doesn't really matter. There is no point in talking about sand, risks, money or technology until we have embraced the cause we stand for. My most important takeaway from the symposium is: seek connection.

RIAN VENS-HAGTING

Understanding problems

Since the financial crisis, ABN AMRO has increasingly viewed its commitment to society and to these organisations and companies from a stakeholder, rather than a shareholder, perspective. This is reflected, for example, in the purpose, which is an important foundation for the day-to-day provision of services: 'Banking for better, for generations to come.' This also aligns with the central question posed in *Laudato Si'*.

According to Richard Kooloos, however, the bank can increase its proficiency in the integral language of these themes:

New problems create new risks and uncertainties. These risks require new solutions and a new way of viewing the future and each other. This affects our function as a bank. Don't run away from climate change problems, such as water, but recognise them and tackle them head on. From the perspective of integral ecology and not just from the perspective of financial risks. Only then will you be able to find sustainable solutions together.

RICHARD KOOLOOS

Taking responsibility

This and the call to action to the financial sector to commit to achieving environmental justice constantly confronts us, as a bank, with fundamental questions.

How do we define the values we choose to engage in shaping the future? And how can we, as ABN AMRO, make the best possible contribution to that just society, in cooperation with all those involved?

The answers to these questions can only be attained when we and all parties involved, regardless of our backgrounds, find sustainable solutions. Not by running away from problems, but by tackling them. And by stepping outside our comfort zone and broadening our horizons.

The symposium showed us that this connection can only be achieved when we take responsibility as stewards of the environment. That is why we work together on new initiatives that enable establishing and strengthening connections between companies and organisations. For example, by connecting new innovative projects with traditional companies that are looking for alternative and sustainable business practices.

Our aim is to learn to look each other in the eye without prejudice, and not be afraid to answer the questions we see in them. Because this is also what the symposium showed us: only when we empower each other can we create a better world.

7.

The Role of Religion in Society

So far, this book has examined how a values-driven dialogue could inspire and broaden the discussion around water and climate change. We have seen how bringing a religious perspective into a traditionally secular arena can help address often overlooked questions around values, worldviews, and existential threat.

In this last chapter we widen our scope and look at the bigger picture, exploring the role of religion in the public sphere and what 'religion as force for good' means. In academic and political circles, many were convinced that modernisation would lead to secularisation and to the decline of religion, the so-called 'secularisation theory'. However, the last decades of the twentieth century provided a massive falsification of this theory. In the last two decades, religion has returned to the public sphere in modern societies. Often this return is associated with violence, like that of 9-11. The return of religion is also associated with refugees and immigrants, and can create a tension between the freedom of religion and the freedom of opinion. Consequently, religion is often a feature of discussions about security in policy circles.

In this chapter, Jan Anthonie Bruijn, President of the Dutch Senate, highlights the Dutch way of respecting freedom of religion and belief in the public sphere. Ernst Hirsch Ballin critically reflects on when to call religion a force for good. Finally, Ecumenical Patriarch Bartholomew accentuates the role religion can play in striving for justice and peace. He stresses the importance of welcoming strangers and refugees into our communities and our hearts.

The Netherlands as pluralist society

PROF. JAN ANTHONIE BRUIJN

PRESIDENT OF THE DUTCH SENATE

We are fond of calling the Senate a *chambre de réflexion*, a room for thorough consideration of proposed laws and government policies. Our democratic system was built on the fundamental freedoms, starting with the freedom of religion that was explicitly acknowledged as early as 1579 with the Union of Utrecht. This has been part of our peace treaties, together with the care for minorities and the responsibility to protect our society against the negative ways in which worldviews and also religions can be used.

The Netherlands is a pluralist society with a secular government. This has helped us over time to develop equal rights for each and every person. The Dutch way of respecting pluralism and secularity allows for the expression of religious diversity, rather than trying to eliminate this from the public sphere. Obviously this has been contested from the beginning and remains so today. That is one reason why our government has appointed a Special Envoy for Religion and Belief, Mr. Jos Douma.

Bringing together diplomats, religious leaders, members of parliament, academic researchers, and leading representatives of NGOs, creates an extraordinary opportunity to reflect on the role of religion in our common search for peace and security.

Religion and security

PROF. ERNST HIRSCH BALLIN

TILBURG UNIVERSITY, FORMER MINISTER OF JUSTICE OF THE NETHERLANDS

The civil wars in countries around the Mediterranean, but also the 'Troubles' in Northern Ireland left many people with the impression that religion is a source of hatred and violence towards people with a different conviction. Religion is thus viewed as a source of insecurity. This view has even gained some popularity among anti-religious opinion makers, as if it were the ultimate proof that religion is a source of intolerance.

Fundamentalists have thus hijacked the public image of religion. Religions are identified with the political movements that abuse religion as a legitimation for pretended supremacy. Recently, various religious fundamentalisms have come to the forefront. What they have in common is their self-understanding as a collectivity that is singular and supreme in its mission. Their identity must not be put on an equal footing with other identities or beliefs, and therefore it is legitimate – or even their historic mission – to reject equality before the law. Ideologies of group supremacy based on a religious or ethnic identifier complete this ill-fated process. They are the 'true believers', the 'true Finns', or whatever. All these 'culture-essentialists' (an expression coined by Andreas Reckwitz) recreate their foundational narratives, either vindicating a religious obligation to impose their will on others, or re-inventing a glorious phase in national history that should serve as the model forever. Collectivised hubris and lust for power can then seduce people to view the other and the other group as the enemy.

Many of us are inclined to respond with the assertion that to us, as believers, religion means non-violence and peace. That may be true, but is that a sufficient response? The question is not what religion means for its followers – the internal perspective – but what it means for others – the external, societal, and political perspective.

The WRR (Scientific Council of Government Policy) has argued in a report[7] (of which I was the project chair and co-author) that security cannot be reduced to defence against intruders:

> "[S]ince the 1980s, and to an even greater extent since the end of the Cold War, there has been a significant increase in concern for human rights and for economic and social development (human security). After all, physical violence is partially explained by the structural violence resulting from disadvantaged social circumstances. National and international security are therefore linked to the security of the society and the individual. [...]
>
> The Brundtland Report already drew attention to the security aspects of environmental problems in 1987. More recently, climate change has pushed the ecological threats to security higher up on the political agenda. With the focus on the potential for climate change to cause conflicts, the concept of security has expanded once again.
>
> The economic growth of countries in Asia, the Pacific region, Latin America, and Africa, the worldwide population growth and the rise of a global urban middle class are driving an enormous increase in demand for energy, water, food, minerals, land, and other natural resources. Safeguarding a steady supply of energy, being able to cope with price fluctuations and reducing vulnerabilities by diversifying and making the transition to renewable energy sources are urgent challenges. [...]

7 Ballin, E. H., Dijstelbloem, H., & de Goede, P. (2020). *Security in an Interconnected World.* Springer Nature.

> These trends will also further increase the migration potential in the world, with more of the people concerned coming from unstable and weak states confronted with protracted internal and regional conflicts, as is now the case in Syria, Iraq, Afghanistan and Libya, or with combinations of conflict, drought and scarcity of raw materials and food, as in Yemen and some Sahel countries. The OECD forecasts that by around 2030 almost half of the world's population will be confronted with the negative effects of the rising sea level and that this will cause more people to seek their fortune elsewhere. Unless mankind takes effective action against climate change, the UN High Commissioner for Refugees has estimated that between 250 million and one billion people will be forced to leave their own countries over the next fifty years."

Security is thus related to the realisation of human rights, in the wide sense that encompasses social, economic, cultural, and environmental rights. These human rights are intimately related to the sustainable development goals. Here, we can discern a different, externally oriented dimension of being religious. Believers who see religion not as a service to themselves but as a service to God and other children of God, out of respect for their dignity and the dignity of the creation, will necessarily prioritise the external perspective of their faith.

This has important implications for the connection between religious self-identification and security. If religion does not erect fences and walls excluding or even confining others, it can become a source of reciprocal security based on reciprocal recognition of fundamental rights. This is what I called in a publication some years ago 'religious citizenship'. The context of life projects, protected and supported by human rights, is in our times a world 'on the move': ongoing migration to the cities brings people together from different origins, with different convictions. Under the twenty-first-century conditions of migration and urbanisation, people move around and co-create their changing social fabric. Many of the oppressed people and – consequently – refugees all over the world have suffered from a lack of protection (or even acceptance) of their citizenship on religious grounds, or supposedly religious grounds used as a guise for power hunger. Room for religious diversity is for them a natural requirement of living together.

What is needed in contemporary society is a mutual willingness of political actors and religious leaders to engage in a dialogue, respectful of each other's responsibilities, aiming at an understanding of how religious freedom and democratic legitimacy can be brought to terms. Religion is a contribution to a vital civil society. It should not be a tranquiliser (as it often has been, in the service of the ruling class). Because it expresses itself in moral categories, it is a more productive critical voice than the superficial bashing of the elite by 'social' media.

That is the good of peace rooted in justice, to which believers must contribute, not only for themselves but equally for people with other beliefs or convictions.

Religion as force for good

ECUMENICAL PATRIARCH BARTHOLOMEW

In our humble opinion, disputing the role of religion in the contemporary world is a thing of the past. It reflects an old-fashioned ideology, which no longer corresponds not only to the essential demands of society, but also to the evident crucial functions of religion both on the personal and the social level. It would be an unfortunate loss for the democratic political system today if religion were ignored. The truth is that neither religious principles nor secular values exist in isolation, but both flourish in creative and constructive dialogue.

We are convinced that it is an inaccurate attitude that religion somehow hampers or menaces progress or nourishes only fundamentalism and division. Nevertheless, this mindset creates a moral vacuum in a society that is shaped solely by economic factors in a globalised world. Religion, however, must neither choose, nor be compelled to retreat from the public space. It must be invited to address social and political issues, to respond to the challenge of human suffering and environmental degradation, and to become involved in sensitive discussions about human rights and against intolerance.

In this perspective, the world of faith can be a powerful ally for engaging in issues of social justice. Arguably, religion can function as one of the most persuasive and transformative forces on earth. Its unique vantage point can contribute today to a more 'human' globalisation, as well as advocating for peace and solidarity and against racism and discrimination, while at the same time championing religious tolerance and human dignity. Very rarely is religion not a defining element of the identity and integrity of a community. This is why religion is the subject of renewed interest and attention in international relations and global politics, directly impacting social values and indirectly influencing state policies. Whether dealing with the environment or peace, poverty or hunger, education or healthcare, there is today an increasing sense of common concern and common responsibility. Indeed, any analysis of our contemporary cultural setting would be incomplete without taking into consideration the presence and impact of religion.

The issues that our world faces are in some ways hardly new. History is replete with examples of violence, cruelty, and atrocities, committed by one group of people against another. However, our present situation is, in at least two ways, quite unprecedented. First, never before has it been possible for one group of human beings to eradicate as many people simultaneously; second, never before has humanity been in a position to destroy so much of the planet environmentally. This predicament presents us with totally new circumstances, which demand of us a radical commitment to reconciliation and peace. The threat to the fabric of human life and the survival of the natural environment make this the overarching priority over all others.

Therefore, as faith communities and religious leaders, it is our vocation and obligation to pursue and proclaim alternative ways to manage human affairs, ways that reject war and violence, and instead recognise and strive for justice and peace. Conflict may be inevitable in our world; but war and violence are certainly not. National tensions may be unavoidable across borders; but tolerance and peaceful coexistence are a categorical imperative. Compassion and change through religion can break the cycle of violence and injustice. In the final analysis, making peace is a matter of individual and institutional choice, as well as of individual and institutional transformation.

Building bridges through encounter and dialogue is a universal human principle. It is also a fundamental Christian mandate, as well as a quintessential part of the DNA of the Ecumenical Patriarchate in its mission to the contemporary world. The Ecumenical Patriarchate has always been convinced of its wider role in the world and of its ecumenical responsibility. Since 1977, we have initiated bilateral interfaith dialogue with the Jewish faith; and since 1986, we have convened several high-level encounters with the Islamic community on such topics as authority and justice, pluralism and coexistence, war and peace, as well as religious tolerance and religious rights. All of these conversations and

gatherings have opened our eyes to the diversity of cultures and religions that comprise our fragmented world. They have also widened and deepened our comprehension of the threat of all forms of discrimination and fundamentalism.

We hear often that our world is in crisis. Yet, the truth is that never before in history have human beings had such a wonderful opportunity to bring so many positive changes to so many people simply through encounter and dialogue. Furthermore, there has never been greater emphasis on tolerance for respective traditions, religious convictions, and cultural diversity.

Diachronically, the image we have for ourselves, for the meaning of our life and for our mission in the world, determines our identity and stance in life. Religions do not have ready answers and solutions for all our problems. Rather, religious faith is a source of existential truths for our relation to others and to creation, for our freedom and happiness; truths opening the dimension of transcendence. From the perspective of religion, the value of a culture or a society cannot be judged by the level of its technological development, of its economic progress, or of its institutional and organisational framework. In fact, these important elements are by no means the essence of a culture. The criterion for the quality and level of a society is the way and the degree of the protection of human freedom and dignity and of the promotion of the culture of solidarity.

Permit us to conclude with a biblical image that underlines the importance of encounter and dialogue in society. Sitting under the shade of the oak trees in Palestine, Abraham received an unexpected visit from three strangers. It is a story recorded in the Book of Genesis, chapter 18. In this biblical narrative, it is refreshing to observe that Abraham does not consider his foreign visitors as posing any danger or threat to his lifestyle or property. Instead, he spontaneously and openly shares with them his friendship and his food, his relationship and his resources. In fact, he extends such gracious and generous hospitality that, in the Orthodox Church, this scene is interpreted and identified as the revelation of God himself.

In images that depict this scene, there is always an empty seat at the table where Abraham is feeding his unknown guests. The question that we leave you with is this: How many foreigners are we willing to seat at our table? We must ask ourselves how many difficult issues are we willing to address together as a society? We must question whether we are willing to open our communities and our hearts to share with foreigners and refugees. And, if we honestly want to effect change – in ourselves and in our world – then we commit to working together and no longer in isolation. The future does not belong to the homo clausus, the 'enclosed man', because he does not care for the future.

Pelet
PeleT

Conclusion: concrete steps into a hopeful future

JAN JORRIT HASSELAAR & ELISABETH IJMKER
AMSTERDAM CENTRE FOR RELIGION & SUSTAINABLE DEVELOPMENT, VU AMSTERDAM

The values-driven approach presented in this book is not just a theoretical exercise about how to stimulate cooperation between science, economy, government, and religion. The aim of this approach is also to formulate practical steps forward on shared water issues in times of climate change. Therefore, signing the covenant of hope is not only sharing good intentions, but also real commitment. Hope without real commitment is empty hope or a naïve invitation to a better world. In the period since the water symposium, the parties that signed the covenant have developed various initiatives for further cooperation. To conclude this book, we present some of these initiatives.

First, in the months after the symposium, the Faculty of Religion and Theology at the Vrije Universiteit Amsterdam integrated themes from the symposium into its curriculum. Several students completed internships or theses on topics related to religion and climate change, and the faculty developed a bachelor course on religion and sustainable development. This course introduces the multifaceted intersections between religion and sustainable development. The course is interdisciplinary and open to all students. The course involved the participation of many partners, including the Netherlands-Indonesia Consortium for Muslim-Christian Relations, Tearfund, Orthodox Church, University of Maastricht, University Groningen, United Nations, Religions for Peace, Van Oord, and ABN AMRO. The participation of several parties has formed a next step following their signing of the covenant of hope.

Second, in the months of the outbreak of corona, the covenant resulted in a cooperation on developing the water strategy of the city of Cape Town, which is in the process of becoming a water sensitive city by 2040. The project focuses on developing community and building trust between city officials and people

in townships. It includes the role religious communities can play in developing a water sensitive city. The project echoes Archbishop Thabo's prayer in this book: 'May we become agents of hope'. Leading partners in the project are the Water and Waste Department of the city of Cape Town, University of Western Cape, Waternet Amsterdam, and Vrije Universiteit Amsterdam.

Third, Vrije Universiteit Amsterdam, together with WINNER (Week of Indonesia - Netherlands Education and Research), developed a follow-up to the Climate Adaptation Summit (CAS2021). This follow-up is called 'A Journey of Hope for a Climate-Resilient Future', and took place as an online event held in March 2021. The aim of this event was to explore forms of cooperation between and within nations to make climate adaptation work for everyone. The event consisted of high- level keynotes, contributions from science and business, and voices from the grassroots. Furthermore, the event explored the prerequisites for successful climate adaptation.

Fourth, as a result of the covenant, the Amsterdam International Water Week 2021 will continue the critical water conversations with a round table on religion and water. The focus of the round table, a conversation with all stakeholders involved, will be on how to further integrate religious communities, leaders, and universal values thematised by religious traditions into developing water policy.

Concluding, we see that water in times of climate change is often surrounded by an atmosphere of fear and apocalypse, water as a threat. By presenting a values-driven approach, we aim to change the narrative and develop a hopeful way forward. In this way forward, working together on shared water issues, can itself become a ritual of cleansing, that enables societal parties, governments and academic disciplines to develop a hopeful future, one in which water can be a source of new life in community building between people, as well as between people and nature. The four initiatives presented above are concrete examples of steps into a hopeful future that are taken in several contexts and for different target groups. These concrete actions show us that hope is within reach. The only thing we have to do is to respond to its call. By responding, we can gradually learn together how to claim its promise and carry into a radically uncertain future that precious torch of hope. To use the words of Amanda Gorman's poem at the inauguration of President Joe Biden:

For there is always light,
if only we're brave enough to see it
If only we're brave enough to be it[8]

It is our hope that we are brave enough to see and be it.

8 *The Guardian*. (2021) 'The Hill We Climb: The Amanda Gorman poem that stole the inauguration show'. Available from: https://www.theguardian.com/us-news/2021/jan/20/amanda-gorman-poem-biden-inauguration-transcript [Accessed 18 March 2021]

alle
doorvaarten

Illustration acknowledgements

City of Cape Town	16-7, 41, 42, 52, 110-1, 144-5
Edwin van Eis	8, 34, 66, 78, 94, 102-3, 121, 150
Carel de Groot	2, 65, 69, 93, 95-6, 118, 134, 146
Jan Jorrit Hasselaar	47, 48, 126
Kadir van Lohuizen	28-9, 30-1, 32-3
Danielle Roeleveld	13, 15, 18, 21, 24, 26, 35, 37, 39, 45, 51, 54, 70, 73, 75, 80, 99, 100, 101, 105, 106, 107, 113, 128, 132, 135, 136, 140, 142, 143
Guus Schoonewille	49
Jos van den Tempel (ABN AMRO)	9, 120, 122, 124, 125, 127, 131
Marieke Wijntjes	91
Kevin Winter	7

About the editors

Jan Jorrit Hasselaar is a theologian and economist. He coordinates the Amsterdam Centre for Religion and Sustainable Development at the Vrije Universiteit. His main work is on theology as a perspective of the good life based on hope in conversation with economics, church, and society at large.

Elisabeth IJmker has a background in public policy and international development. She works on religion and societal challenges at the Vrije Universiteit, which she combines with a political career as local representative in the city council of Amsterdam.